ENERGY
Demand vs. Supply

Edited by Diana Reische

THE REFERENCE SHELF
Volume 47 Number 5

THE H. W. WILSON COMPANY • New York • 1975

THE REFERENCE SHELF

The books in this series contain reprints of articles, excerpts from books, and addresses on current issues and social trends in the United States and other countries. There are six separately bound numbers in each volume, all of which are generally published in the same calendar year. One number is a collection of recent speeches; each of the others is devoted to a single subject and gives background information and discussion from various points of view, concluding with a comprehensive bibliography. Books in the series may be purchased individually or on subscription.

PRINTED IN THE UNITED STATES OF AMERICA

Library of Congress Cataloging in Publication Data
Main entry under title:

Energy :demand vs. supply.

(The Reference shelf ; v. 47, no. 5)
Bibliography: p.
SUMMARY: A compilation of articles discussing the sources of energy and their alternatives, the emergence of the energy crisis, the need for conservation and efficient use of available energy, and the economic and political considerations influencing United States policy,
I. Reische, Diana L. II. Series.
TJ163.2.E47 333.7 75-28149
ISBN 0-8242-0573-1

PREFACE

Energy runs the vast industrialized civilization that Americans have taken for granted for several decades. It powers not just the machinery and the factories, not just the transportation, communications, and utilities systems, but also the appliances that provide the ordinary comforts of daily living—television, air conditioners, phonographs, toasters, dishwashers, hair dryers.

We have become so accustomed to the conveniences provided by abundant and seemingly unlimited energy that we have taken them for granted. We have expected that no matter how many air conditioners were plugged in on a given day, the local electric company could supply enough current to run them all. We have expected that no matter how many cars there were on the road, gasoline would be available at a nearby pump whenever the gas gauge indicated "empty." Furthermore, people in the United States could expect that the prices paid for various forms of energy would be relatively low in comparison to costs elsewhere in the world.

Throughout our history an ever-increasing demand for energy was looked upon as beneficial. Supplies were abundant and varied. Swift-flowing rivers were harnessed to hydroelectric plants; vast coal fields and oil pools were exploited. And the economy boomed. Each new industrial surge brought with it the need for more energy and each increase in energy supply spurred industrial growth.

Without really noticing, the United States had become, by 1975, an energy glutton. With only about 6 percent of the world's population, this nation used one third of all the energy consumed globally. Much of it, unfortunately, was simply wasted in inefficient machinery, oversized cars, poorly insulated buildings, and a hundred other energy-devouring ways.

Warnings that such a situation could not continue indefinitely became more frequent in the 1960s, and by the early 1970s the phrase *energy crisis* had slipped into the national vocabulary. Today there can be few who have not personally experienced a gasoline waiting line, a skyrocketing fuel bill, or a school or office several degrees cooler in winter because of rising fuel costs.

For a while, many people placed the blame for the crisis on the Arabs, who, as owners of most of the world's known oil reserves, boosted prices sharply. But as more information became available, as new problems of energy supplies arose, the realization grew that the crisis would not go away even if the oil-producing countries decided tomorrow to drop prices by half.

If the crisis is genuine—and most analysts consider it to be very real and very immediate—what solutions exist? Which solutions offer the best long-run advantages?

This book offers a wide-ranging view into both the development of the energy problem and the possible ways of dealing with it. Unfortunately, there do not appear to be definitive, immediate solutions. Modern economies require great amounts of energy. Yet existing sources are already strained. The readily available alternatives to oil—such as coal—have the undesirable side effects of pollution and environmental spoilage.

More exotic forms of energy such as solar power and nuclear energy are many years away from wide-scale production. Each requires enormous advances in technology to permit mass production at reasonable costs.

Thus many analysts believe that there are at least two separate time-frames to consider in looking at the energy future of the United States. In the long run, it seems possible that some way will be found to utilize some of the now exotic energy sources economically.

It is the short term—say the next ten to twenty years—that appears to many observers to be the most difficult period in matching energy supplies with energy needs. During this period, and perhaps even farther into the future, costs will probably continue to rise. The day of cheap energy may be gone for the foreseeable future.

Whether now or in the long run, it is vital that demand be reduced, or at least the *growth* in demand be reduced, by making more efficient use of whatever energy is consumed. Such a reduction will require cars with better gas mileage, air conditioners that draw less power, buildings that are better insulated, machines that run more efficiently.

This book is divided into five sections that examine separate aspects of the energy dilemma. The first is an overview of how and why the energy problem emerged. The opening section looks, too, at the technical side and language of energy: what it is, how it is measured, how it is transmitted.

The second and third sections explore in depth the fossil fuels—oil, natural gas, and coal—which today provide 95 percent of all energy needs in the United States. These are the primary energy sources available in the next two decades.

A fourth section explores the alternatives to fossil fuels. These range from the familiar windmill to the exotic, such as nuclear fusion and fuel cells. Throughout the discussion of alternatives, however, runs the theme that even while developing new energy resources it is imperative to conserve existing sources and to use with greater efficiency any energy that is consumed.

A final section explores some of the economic and political choices now available, describing the new federal energy agency, and offering several (sometimes conflicting) suggestions on what the United States should do next to cope with the energy dilemma. Again, virtually every author argues that more careful use is imperative.

The editor would like to thank the authors and publishers for permission to reprint the excerpts in this volume.

DIANA REISCHE

August 1975

CONTENTS

IV. Alternatives to Fossil Fuels

V. The Energy Options

I. ENERGY AND POWER

EDITOR'S INTRODUCTION

The energy crisis—how did it develop and what is its future? These are fair and pertinent questions when anyone surveys the US energy situation in the mid-1970s. In this section several authors offer separate routes into the maze that is the energy dilemma.

Each tries to illuminate in his or her own way the complicated nature of the energy situation. Vital decisions on energy use and allocation occur in both the political and the economic spheres. The history of energy growth in the United States has been in many ways the history of the growth of the American economy itself. If that energy growth stops or slows down, what happens to the economy?

The problems are immediate and affect our lives, our jobs, our homes. How the problems evolved, and whether they might have been avoided, are touched on in several articles. There is no single villain. Shortsightedness seems to have been as much a reason as greed for the current situation.

For decades the US economy boomed forward on ever-increasing amounts of cheap energy. Where once coal was the "big" fuel, the United States in this century shifted to natural gas and petroleum as primary energy sources. The few voices warning that these fossil fuels were nonrenewable and that once used were gone forever were drowned out by the chorus of optimism. It has been almost an article of faith that geologists and engineers would pull another rabbit out of the energy hat when needed.

Technology and life styles, too, aggravated the difficulties. Since fuel was cheap and abundant, industry had no incentive to design cars that consumed less gas, or to make buildings that were better insulated, or to invent machines that could run longer on less fuel. The spread of homes and businesses to outlying suburbs increased fuel consumption of cars and buses.

The introductory article in this section is a historical overview of the development of energy utilization in the United States. The author, S. David Freeman, is an acknowledged expert who directed the recently completed Ford Foundation study on energy discussed in Section V ("A Time to Choose").

Next, a short energy "dictionary" is offered as a guidepost for readers unfamiliar with the technical jargon of energy and power.

The following article, "Understanding Energy," explains just what energy is. The author, physicist John Holdren, offers this rough distinction: energy is the capacity to do work, while power is the rate at which work is done, or equivalently, the rate at which energy is consumed. Mr. Holdren introduces such concepts as energy conversion, the laws of thermodynamics, and the fact that unlike many resources, energy cannot be recycled.

Next are two articles prepared by the Council on Environmental Quality: the first makes projections about how the present use patterns of the various fossil fuels will shift in the next two decades; the second describes the various ways energy consumption degrades or pollutes the environment. The environmental problems range from air pollution by automobiles to land spoilage by surface mining. The conflict between concern for the quality of the environment and the wish to develop additional sources of energy is a continuing theme in this book and occurs in articles on many of the specific fuels: coal, because of sulfur emissions; oil, because of tanker spills; nuclear power, because of safety problems; shale oil, because of threats to the water table; and so on. There are, in fact, environmental costs for nearly all fuel sources.

Who in the government is supposed to plan and worry about these things? An article from *Congressional Digest* lists the tangle of federal agencies, departments, and commissions that have been charged with responsibility for directing various energy programs.

THE AGE OF ENERGY[1]

As we suddenly discovered in the winter of 1973–74, our way of life is completely dependent on a reliable and adequate supply of energy. The fuel and electricity that is derived from oil, natural gas, coal, hydropower, and the atom have steadily taken over much of our daily work. Energy cooks, washes, dries, vacuums, heats, cools, and illuminates most American homes. It drives us back and forth to our jobs, takes us on vacation, and brings us most of our entertainment through TV, radio and movies. In the factories and commercial establishments where most of our energy is used, energy greatly assists the worker and enhances his productivity by powering machines and automated equipment. The fuels and electric power used in the United States provide every American with the energy equivalent of as many as two hundred full-time personal servants. Those people living in poverty consume less energy and possess relatively few such servants, but they too seek to join the high-consuming majority.

The big blackout in New York City and most of the Northeast late in the afternoon of November 11, 1965, first brought home to many Americans our near total dependence on energy. Those who experienced the shock and fright of being trapped in elevators, in subways or in their offices, unable to get home, have not soon forgotten it. No one knew what caused the blackout. President Johnson heard the news on the radio at his Texas ranch, and his first reaction was to call Defense Secretary Robert McNamara and be sure that our strategic defense systems were alert and operable. As the

[1] From *Energy, the New Era* by S. David Freeman, an engineer and lawyer who was from 1967 to 1971 the chairman of the President's energy policy staff. Walker. '74. p 14-23. Published by Walker & Company, Inc. New York, N.Y. Copyright © 1974 by The Twentieth Century Fund. Reprinted by permission.

blackout affecting thirty million people persisted through the night, the President asked the Federal Power Commission [FPC], the agency that regulates the electric power industry, to investigate. The FPC was as stunned and unprepared as the average New Yorker, and the utilities affected were so unprepared that the lights even went out on the panels controlling the power plants themselves. The blackout of the Northeast could have been a major tragedy instead of an expensive inconvenience. Only a bright full moon prevented serious loss of life as airplanes landed without light. Even so, property damage from food spoiling in freezers, machines stopping without warning, computer time lost and a host of other causes was estimated in the millions of dollars. Although the statistics showed a measurable increase in the birthrate nine months later, most people did not find the blackout highly enjoyable.

The electric power companies since have been able to prevent recurrences of regionwide blackouts. But power shortages in one region or another have become regular events called brownouts. Voltage is reduced, which, within limits, has only minor effects such as reducing the quality of television pictures. If the brownout is severe, power is rationed and certain industrial customers may be cut off. If industrial customers cannot be cut off, demand is reduced by cutting off an entire area for an hour or two. The cut is rotated among areas so that inconvenience is spread more or less equitably. The brownout technique is used to reduce demand so that everyone's service doesn't black out at once.

By 1972, the summer power shortage was accompanied by a gasoline shortage. And then these were accompanied by the winter fuel shortages, which first hit the Midwest in the winter of 1972. A year later, the nationwide energy crisis developed. Americans now receive year-round, inconvenient reminders of how much their comfort and essential services depend on energy.

When measured from a long-term perspective, modern society's current dependence on commercial energy has developed quite swiftly. As recently as 1850 people in the United States still obtained as much as two thirds of their energy from human muscle power and draft animals. Wood was the main fuel, and almost all households burned it. For industry, the water wheel furnished power directly and windmills supplied most of the mechanical horsepower. The amount of energy consumed was quite small. Indeed, all of the commercial energy consumed by mankind prior to 1900 probably would not equal this year's consumption alone.

The Industrial Revolution ushered in what may be called the Age of Energy. It got under way in this country in the 1860s, for the most part, and was marked by the increased use of the steam engine, which required great amounts of fuel, mainly coal. Lying dormant in the earth's crust, virtually untapped by mankind, were the coal, oil, natural gas and other sources of commercial energy needed to fuel the new technologies as fast as they were being developed. The chemical interactions and work of nature over 100 million years were required to form fern fossils into these hydrocarbons. Thus, once used, they could not be replaced for at least another 100 million years. This is a sharp reminder that even a seemingly ample 300-year supply of fuel would last but a mere ten seconds in the long year of nature's complex and creative manufacturing process.

Coal was an important source of energy in the US Industrial Revolution. Its major uses were for generating steam and in making iron and steel. Coal, of course, had been used in Europe for centuries and many small coal mines had existed in America since colonial times. But as long as firewood was plentiful and mining machines and transportation were limited, coal was not used extensively. Until the early nineteenth century, even steam engines, river boats and railway locomotives were fueled by wood.

After 1865, coal came into its own. The remarkable and unparalleled growth in the production of coal from 20 million tons in 1860 to 500 million tons in 1910 powered the industrialization of the nation. During this transformation and redirection of a once agricultural economy, coal displaced firewood in most homes and industries. If we combine bituminous coal (which was used mainly in industry) and anthracite coal (which was preferred in households), coal had captured 90 percent of the energy market by 1910.

Coal ushered in a new era in American civilization and radically changed people's life styles. People moved into urban America to be near the factories that offered jobs. The coal-fired steam engine made possible large-scale operations that were much more economical than small plants, and big plants steadily replaced the small-scale, decentralized and subdivided manufacturing process. The steam engine not only brought economies of scale, it also became the epitome of progress.

This large-scale exploitation of America's coal seams marked a new turn in our civilization, the beginning of intensive exploitation of an asset of limited life in contrast to the renewable wood, wind, and water power of the past. As we drew on this irreplaceable resource capital, another turning point just as significant in human terms occurred, which Louis Mumford has described:

> Now, the sudden accession of capital in the form of these vast coal fields put mankind in a fever of exploitation: coal and iron were the pivots upon which the other functions of society revolved . . . The animus of mining infected the entire economic and social organism: this dominant mode of exploitation became the pattern for subordinate forms of industry . . . The psychological results of carboniferous capitalism—the lowered morals, the expectation of getting something for nothing, the disregard for a balanced mode of production and consumption, the habituation to wreckage and debris as part of the normal human environment—all these results were plainly mischievous.

The oil industry also had its beginning just as the nation started to industrialize. The world's first commercial oil well was drilled by E. L. Drake in 1859 near Titusville, Pennsylvania. Interestingly enough, the first market for oil was as a medicine and the first mass market was as kerosene, a fuel for lamp light which replaced coal and whale oil.

Oil also served as a lubricant essential to the smooth functioning of the new industrial machines. Within a few years, oil captured most of the US lighting and lubricating markets and was being exported to Western Europe. During the Civil War, oil exports helped preserve the Union by replacing southern cotton as an exchange for imports vital to the war effort. For the next century, the US oil industry dominated the world oil scene.

As Americans began to shift to the new ways of industrial life after the Civil War, John D. Rockefeller made his great fortune in oil. He consolidated his interests in a ruthless manner which left an indelible imprint of concentrated economic power on the entire oil business. His operation, even though quite efficient and most enterprising, stamped the oil industry as a strike-it-rich, public-be-damned monopoly. But the early capitalist period also saw considerable growth and improvement in the technology of the industry. Refinery capacity initially was centered in Pittsburgh but it soon spread to major cities of the East Coast and expanded rapidly. Marketing and distribution systems benefited from the introduction of such innovations as pipelines, first made of wood and then of iron and steel. The rail network, using rudimentary tank cars—again, initially made of wood—was extended into the oil fields.

Even with all of the manipulations of the marketplace and the secret, anticompetitive dealings, the oil industry itself actually got off to a slow start as a major source of energy. In its first forty years, the period 1860–1900, a total of only one billion barrels was produced—the equivalent of about three months of current production. Even so, by the turn of the century, the consensus was that we were "running out of oil." This attitude changed dramatically on January 10, 1901, when "big oil" first gushed from the earth at a spot in southeast Texas called Spindletop. The next year, Henry Ford organized the company that ignited a gigantic new market for oil. The mass production of Model-T Fords and the first successful aircraft flight at

Kitty Hawk by the Wright brothers in 1903, launched America into the age of the motor vehicle and the airplane, two forms of convenient, comfortable and swift transportation that soon began to dominate our lives.

The remarkable period of technological innovation and economic expansion in the United States also witnessed the beginning of the electric power industry. Thanks to the ingenious inventions of Thomas Edison in the field of electricity, the first central, steam electric plant in the world began operating in New York City in 1882. The kerosene lamp and the gaslight were thus quickly doomed to be replaced by the brighter, more convenient and apparently cleaner electric light bulb. Actually it was the electric motor much more than the light bulb that revealed the flexibility and versatility of electricity as a source of commercial energy. The electric motor spurred the growth of American industry by increasing efficiency and productivity in the country's manufacturing plants. Industry adopted electric motors with great speed, and by 1920 they accounted for more than one half of total installed horsepower in factories, compared with less than 5 percent in 1900. These early uses of electricity were, of course, only the beginning of an endless array of appliances and conveniences and important new industrial uses.

Thus within a short period around the turn of the century, America witnessed the emergence of the major segments of what has become today's energy industry and the complex, fuel-hungry technology powered by these forms of energy. Once under way, the energy industry advanced quickly, and irresistibly. The world of the machine had arrived and with it the joys of unbounded material success, and the sorrows of lost values accompanied every clang, bang and roar.

Coal was still king in the production of energy in the early twentieth century, but events began to take shape in peacetime and in war which led to the rapid conversion to oil. Winston Churchill, as first lord of the [British] Admiralty just before World War I, was impressed with the vital role oil could play in developing a navy that was faster and able to stay at sea longer than coal-fired ships. As it happened, war broke out in Europe before his plans for exploration in the oil-rich Middle East could be implemented. The strategic importance of oil in World War I was established in other ways. By cutting off German oil supplies and war material, the Allied blockade hampered the newly mechanized German armies and severely handicapped needed war production.

After the war the United States, impressed with the importance of oil, once again feared that its own oil sources had already been exhausted. In May, 1920, Senator James Phelan of California suggested the establishment of a federally owned company to search for oil abroad. In 1922, these fears were alleviated by major new discoveries in the United States and Venezuela. But the oil-rich Middle East was the real prize that interested all the industrialized nations and the international oil companies which dominated the market. In the post-World War I period the British and French became very active in trying to obtain concessions to explore for oil in the Middle East. In 1927 the vast oil resources in that part of the world began gushing forth in Iran. The largest US companies, such as Standard of New Jersey, Mobil, and Texaco, as well as British and French interests gained a share of what became by far the major source of oil for the entire world. The Middle East today contains 63 percent of the earth's known reserves.

Spurred by the new discoveries of oil and popularity of the automobile and other new oil-consuming machines, the Age of Energy steadily shifted the focus of consumption from coal to petroleum. Motor vehicles and airplanes provided the major market for oil, but other markets also opened as railroads and homes switched to the cleaner and more efficient liquid fuel. This shift is still continuing as new discoveries in the Middle East and more recently in northern Africa, the North Sea and Alaska attempt to keep pace with the remarkable growth in the consumption of oil.

Natural gas was another moving force in the transition from coal to petroleum during the past three decades. Natural gas is a relatively new source of energy but the local gas companies that sell it date back to 1865, when gas manufactured from coal was first sold in significant volumes. Some companies are even older, but around 1865 improved technology gave birth to the modern gas industry by greatly reducing the cost of manufacturing gas from coal. After World War II they simply switched to natural gas because it was much cheaper.

Natural gas was at first considered an unwanted by-product in the search for oil. Thus, much of the gas was flared as the oil was produced, or if found in a separate well, it was shut in for lack of market. The early usage of gas was confined to the immediate vicinity of the fields and to a few nearby industries because of the lack of technology to transport it very far. As pipeline technology improved, more and more residences and factories switched to natural gas, which was cleaner, easier to use and cheaper than coal.

An important breakthrough came in the 1930s: the development of large-diameter pipelines capable of withstanding the high pressures necessary to transport natural gas long distances. Pipelines could now economically carry gas many hundreds of miles. A few of these early pipelines brought gas from the fields in the Southwest to the markets of Chicago, Minneapolis and Detroit.

World War II delayed the major expansion of the natural gas industry into other markets. But in the postwar period, natural gas was delivered throughout the nation, and it rapidly replaced coal in homes and industry in the 1950s and 1960s. In fact, natural gas has supplied more than half the growth in total energy use in the United States since World War II. Today, fully one-half the nation's 63 million homes are heated by natural gas and most of the remainder by oil.

The most marked change in the American fuel picture over the past twenty years has been the decline of coal. In 1950 coal still provided almost 40 percent of this country's total energy consumption and was of major importance in both transport and industry, though already declining as a residential fuel. By 1960, coal's share had fallen to 23 percent and has since declined to 17 percent in 1972. Coal's main end use now is as fuel for the electric generating industry. It has lost its place to oil and natural gas, which now account for 75 percent of the United States' total energy supply.

Several factors lie behind these considerable shifts in fuel supplies. Changing technology, such as the dieselization of the railways which ended coal's role in the transport sector, have been responsible for some of the changes. Also, the fact that coal is dirty and inconvenient to store and handle has discouraged its use, particularly as a residential fuel. Recent environmental concerns have further contributed to coal's unacceptability.

In comparison to coal, oil and particularly natural gas, are clean fuels and are well adapted for use in some of the major energy end uses—space heating and industrial processes. The phenomenal increase in the use of natural gas is also due to its relative cheapness, although there are great geographic variations in its price compared with other fuels. At the beginning of the 1950s, gas was one-third the price of coal, its cheapest competitor. Thereafter, the price of natural gas rose steadily until, by the end of the 1960s, it was only slightly cheaper than coal in many markets. But the environmental advantages of natural gas were sufficient to make it favored over coal—indeed, so favored that it cannot satisfy the demand.

The development of low-cost transportation is an important element in the growth of our use of all forms of energy. Natural gas has, of course, benefited from low-cost, large-diameter pipeline transport. Coal, though heavy and bulky, can still be shipped several hundred miles to market by train. Recently the unit train, which carries coal non-stop from mine to power plant, has greatly enhanced coal's economic usefulness and caused it to grow in recent years as a power generating fuel. Similarly, the new supertankers, capable of transporting 300,000 tons or more of oil, make it possible for Middle Eastern oil to move halfway around the world at relatively low cost. Extra-high-voltage electric power lines now enable us to locate generating plants outside of congested cities and near supplies of fuel and cooling water. But transportation, though it has created virtually a worldwide market, a market that keeps growing by leaps and bounds, cannot add to the finite supply of fossil fuels.

The commercial atom thus has entered the scene at a very opportune time. The power of the atom was, of course, first unleashed by the United States in the World War II Manhattan Project and then delivered against the Japanese cities of Hiroshima and Nagasaki. Out of a combination of guilt, pride, technological wizardry, and governmental foresight, we have since harnessed this power of the atom for the peaceful purpose of generating electricity.

The Atomic Energy Commission was established by the Congress in 1946 and given responsibility for both the development and regulation of atomic power. After a $3-billion, 25-year effort, the commercial atom came of practical and working age in the 1960s. But debate still rages over whether its poisonous byproducts can be successfully controlled. Despite a good record to date there are still many unanswered questions concerning the future of atomic energy. Nevertheless, we are relying on the atom to supply most of the future growth in electric power supply, the only form in which we have thus far learned to harness atomic energy. And so the transition from wood to coal and then to petroleum as the main source of commercial energy now appears to lead to the atom, but this is probably not the last link in the chain. One day, we will finally turn directly to our planet's ultimate source of energy, the sun.

The energy industry in America developed swiftly because it was innovative and dynamic. Abundant sources of new fuels have been opened up even while supplies of traditional fuels were still plentiful. But the world has never experienced energy consumption on the present scale, and by the early 1970s the energy industry had become a victim of its own success. It tried to grow faster than its capacity in a time of environmental and political constraints at home and abroad. As a result, our high-energy civilization is in trouble.

ENERGY DICTIONARY [2]

Energy—What Is It?

Something which gives us the capability to do mechanical work or to produce a change in temperature (that is, to heat or to cool).

What Form Does It Take?

Energy can take many forms, such as mechanical motion (. . . "kinetic energy"), temperature difference between two objects ("heat energy"), and the flow of electricity ("electric energy"). "Potential energy" is mechanically stored energy, as in the tension in a spring, or water stored behind a dam, or chemically stored energy, as in a fuel.

What Is the Relationship Between Energy and Power?

Power is the rate at which energy is used or the rate at which work is done.

Where Does Our Energy Come From?

In our civilization today, most energy is derived by burning fossil fuels (release of stored chemical energy). The fossil fuels are petroleum, natural gas, coal (anthracite, bituminous, lignite, and brown coal), oil shale, and tar sands. They are the products of biological processes modified over

[2] Adapted from the pamphlet *Citizen Action Guide to Energy Conservation*, prepared by the Citizen's Advisory Committee on Environmental Quality. '73. Supt. of Docs. Washington, D.C. 20402.

millions of years by geological processes. A small amount of energy is also supplied by flow of water, nuclear processes, wood, and the muscle of man and animals.

How Do We Specify Amounts of Energy?

The quantity of energy can be expressed by a variety of equivalent units which apply to the mechanical, heat, or electrical forms of energy. The most commonly used units of measurement are:

☐ *British Thermal Unit (BTU)*

One BTU is the energy required to increase the temperature of 1 pound of water by 1 degree Fahrenheit. For example, it takes 300 BTUs to heat 1 quart of tap water to boiling. How is that figured? One quart weighs 2 pounds. Assume tap water temperature is 62 degrees Fahrenheit. Boiling point is 212 degrees Fahrenheit, so needed temperature rise is 150 degrees Fahrenheit. The BTUs needed are 2 pounds $\times$ 150 degrees Fahrenheit = 300 BTUs.

☐ *Kilowatt (1000 watts)*

A unit of electrical power indicating the rate at which electrical energy is being produced or being consumed.

☐ *Kilowatt-hour (1000 watt-hours)*

A unit of electrical energy equal to the energy delivered by the flow of 1 kilowatt of electrical power for 1 hour. For example, a 100-watt bulb burning for 10 hours will consume 1 kilowatt-hour of energy.

Energy Efficiency

The amount of useful work or product divided by the fuel or energy input. For example, in electrical generation it is the amount of electricity produced per unit of fuel consumed.

Commonly Used Abbreviations

bbls—barrels (a barrel contains 42 gallons)
kw—kilowatt

kwh—kilowatt-hour
mcf—1,000 cubic feet (of gas)
mw—megawatt, 1 million watts
BTU—British Thermal Unit
therm—a unit of energy used for natural gas equal to 100,000 BTU

Energy Conversion Table

To Convert From	*To*	*Multiply By*
kilowatt-hour	BTU	3,412.8
1 ton bituminous coal	BTU	26,200,000
1 bbl crude oil	BTU	5,600,000
1 bbl residual oil (No. 5)	BTU	6,290,000
1 gallon gasoline	BTU	125,000
1 gallon No. 2 fuel oil	BTU	138,800
1 cubic foot natural gas	BTU	1,031
1 mcf natural gas	BTU	1,031,000
1 therm natural gas	BTU	100,000
1 BTU	kwh	0.000293

UNDERSTANDING ENERGY [3]

Ask a scientist to define energy and you are likely to get a lecture rather than a simple answer. For although energy plays a dominant role in our science, our society, and the processes of life itself, it is readily described only in terms of what it does, rather than what it is. Nobel laureate Richard Feynman, in his celebrated *Lectures on Physics,* calls energy "a numerical quantity which does not change when something happens." His definition refers to the law of conservation of energy, which says that energy is neither created nor destroyed but only changes in form. For most of our purposes here, we shall settle for a less comprehensive but more practical definition: "Energy is the capacity to do work."

[3] From Chapter 1 of *Energy,* by John Holdren and Philip Herrera. Sierra. '71. p 14-24. Copyright © 1971 by Sierra Club. Used with permission. John Holdren, author of this chapter, is a physicist at the Lawrence Radiation Laboratory.

The work it does is the operation of the biosphere and the maintenance of agricultural and industrial civilization. Radiant energy flowing from the sun warms the earth to life-sustaining temperatures, drives the hydrological cycle of evaporation and precipitation, powers the winds and ocean currents, and is captured by photosynthesis to fuel the earth's biota. Of the solar energy reaching the surface of the earth, roughly one tenth of 1 percent is utilized by photosynthesis—two thirds on land and one third in the seas. Of this amount, in turn, the metabolic requirements of man and his livestock account for roughly 1 percent. In a system of some millions of species, of course, the appropriation of a percent of the biological energy flow to serve the needs of only one species is a remarkable feat. Ecologists find this situation disturbing as well as remarkable, because they know that the stability of ecological systems, including those that support mankind, is related to the balanced flow of energy through the widest variety of biological pathways.

Man's use of energy far exceeds his metabolic requirements, however. Historically, his population and his material wealth have grown in step with his ability to harness inanimate energy—first as wood and wind, later as falling water and fossil fuels, finally as the energy of the nucleus. The first four categories represent stored solar energy in different forms. The fifth, nuclear energy, originated when the constituents that eventually became our sun and its planets were fused together from elemental hydrogen in more distant stars. Today, man is mobilizing and consuming energy from these various stored sources at a rate some fifteen times that of his metabolic consumption, and three times that of the combined metabolic consumption of him and his livestock. It is this prodigious use of inanimate energy—its history, the nature of the fuels and the technologies that sustain it, and the implications of present trends—which will concern us here.

We shall find that formulating a rational strategy for energy use requires definitive analysis of a host of questions that have only recently begun to be taken seriously. Is personal well-being—"standard of living," if you will—as directly related to per capita energy consumption as the energy promoters would have us believe? (Obviously there is a connection, up to a point, but the United States as a whole may be well beyond it.) What fraction of today's spiraling energy demand is actually being used to satisfy the undisputed needs of the poor, whose plight is so often held out as an excuse for technological exercises of little relevance to them? How much of the legitimate demand arising from a certain unavoidable amount of population growth, and from the aspirations of the impoverished to a decent existence, can be met by cutting back on frivolous and wasteful uses of energy by the affluent, rather than by increasing production?

In a hypothetical world, free of the constraints of biology and thermodynamics, such thorny, socioeconomic questions might not have to be asked at all; the energy problem would be reduced to the technical details of meeting any demand that happened to materialize. Unfortunately, we do not live in such a world. Energy is not merely the prime mover of technology; it is also a central ingredient in man's impact on his environment. No means of supplying energy is without liabilities, and no form of its consumption is without consequence to the ecosystems that support us.

Thus, burning fossil fuels pollutes the air, defaces the landscape, and depletes a resource ultimately needed for other uses—lubrication, synthesis, perhaps protein culture. Fission reactors generate a burden of radioactive wastes which must be infallibly contained, unerringly transported, and indefinitely interred. Hydroelectric sites are in limited supply, and exploiting them has esthetic and ecological drawbacks. Solar energy is unevenly distributed in space and time, dilute, and correspondingly expensive to harness. Controlled thermonuclear fusion has not yet been conclusively demonstrated to be technologically feasible, nor can anyone say with assurance what it will cost when we get it.

This list does not exhaust the possibilities for energy sources, nor does it include all the defects of the ones named. But it serves to suggest that there are no easy solutions to the so-called energy crisis. The details that follow will not change this conclusion.

Thermodynamics and Terminology

Understanding energy problems in depth requires a passing acquaintance with the principles of thermodynamics and with the terminology and units of measurement associated with energy and power. First of all, energy and power are not the same thing. If energy is defined as the capacity to do work, then power is the rate at which work is done or, equivalently, the rate at which energy is consumed.

Unfortunately, these simple definitions can get us into difficulty. The law of conservation of energy, also called the first law of thermodynamics, says that energy cannot actually be consumed at all. Thus, when we say energy is consumed, we really mean it is changed in form. The second law of thermodynamics, in turn, says that such changes proceed in such a way as to reduce, on the whole, the availability of energy to do work. In other words, the usefulness of energy is effectively consumed, even though the energy is still present in one form or another. Energy that has been changed in form with a reduction in usefulness—that is, capacity to do work—is said to have been *degraded.* For example, in doing the work that makes light, most of the electrical energy passing through a light bulb is degraded to heat. The light energy itself may do work and be degraded to heat upon being absorbed.

Energy can also be transformed in the opposite direction, from less useful to more useful form, but only at the expense of degrading a fraction of the initial amount of energy present. Such processes are called *energy conversion,* and the generation of electricity is one example. Here, the useful heat energy released, say, in the burning of fossil fuel is partly converted into electrical energy of higher usefulness and partly degraded into relatively useless low-temperature heat. (The words "temperature" and "heat" are not interchangeable: heat is a form of energy, and temperature is a measure of the usefulness of that heat. The higher the temperature, the more useful the heat.)

The real message of the second law of thermodynamics, then, is that energy, unlike most resources, cannot be recycled. The total availability, or capacity to do work, in a given amount of energy can be used only once. Even in energy conversion, when all aspects of the process are considered, the total availability of the energy involved has gone irretrievably downhill. Thus the laws of thermodynamics have sometimes been stated this way: the first law says you can't win; the second law says you can't break even and you can't get out of the game.

Power has a broader meaning than simply the rate at which work is done. It is the rate at which energy is processed, where processing can mean conversion, degradation, the performance of useful work, or even simply transmission. It is helpful to write this definition as an equation:

$$\text{power} = \frac{\text{amount of energy processed}}{\text{length of time during which processing occurs}},$$

or, rearranged more tersely, energy = power $\times$ time. Note that energy is the more fundamental of the two quantities. If we have a gallon of gasoline, we have a specified amount of energy; but the power we produce when we burn it can be anything, depending on how *fast* we do so. The gallon might produce 50 horsepower for half an hour in your automobile, or 60,000 horsepower for a second and a half in a Boeing 747.

Unfortunately for laymen and specialists alike, the wealth of material that has been written concerning energy and power is cluttered with a confusing array of units for the measurement of these quantities—for example, British Thermal Units (BTU) for the energy content of fuels, kilocalories (kcal, or cal) for the energy content of foods, kilowatt-hours for amounts of electrical energy. However, one set of units can as well be used for all applications, and relationships are made much clearer by doing so. We shall use the kilowatt-hour as our basic unit for measuring energy. Since it contains a unit of power (kilowatt) and a unit of time (hour), its use as the fundamental unit of energy serves as a reminder of the relation connecting these three quantities. This all-purpose unit is usually denoted *kilowatt-hour (thermal),* abbreviated kwht, to distinguish it from the more specialized unit of electrical energy, the kilowatt-hour (electrical), abbreviated kwhe. The corresponding units of power, kilowatt (thermal) and kilowatt (electrical), are abbreviated kwt and kwe.

Trends in Energy Use

The use of energy in the United States during the past two decades and indeed the past century has been characterized by three central features: enormous growth, major shifts in the relative importance of competing sources, and significant changes in the patterns of consumption (most notably the increasing use of electricity). These features have been tightly interrelated: new uses for energy have stimulated growth and, in turn, been stimulated by it; rising demand has prompted innovation to provide a matching supply, and innovation and competition have kept costs falling, which stimulated still further increases in demand. It has been taken as an article of faith in the energy industries that all these factors will continue to operate in the future as they have in the past—that competing new technologies will drive the cost of energy ever lower, and that the economy will respond by using even more of it per capita. Actually, there are good reasons for believing this will not be the case, and—at least in the United States—there are compelling reasons for working to prevent it. These reasons will unfold as we examine the energy picture in detail. A logical starting point is a closer look at the trends that have been operative up to the present.

The consumption of inanimate energy in the United States during the period 1880 to 1969, broken down by sources, is shown in Figure 1. Two surges of growth are evident, separated by a slack period between 1920 and 1940. During the first surge, from 1880 to 1920, the average rate of increase in consumption was 3.5 percent per year, which corresponds to a doubling time of twenty years. In the second surge, consumption doubled between 1940 and 1965: the rate of growth itself has been increasing since then. The 5.1 percent increase observed between 1968 and 1969 would lead, if it persisted, to additional doublings of US energy consumption every fourteen years. As is shown in Figure 1, total consumption in 1969 was over 19 trillion kwht, or almost thirteen times the 1880 figure. This 1969 US consumption, incidentally, amounted to just over one third of the energy consumed in the entire world during that year.

Almost as remarkable as this phenomenal growth itself is the shifting composition of energy sources that sustained it. In 1880, more than half the total was supplied by fuel wood and nearly all the rest by coal. By 1900, fuel wood's share had dropped to 20 percent; oil, natural gas, and hydropower had appeared on the scene to claim 10 percent of the market among them; and coal accounted for the remaining 70 percent. Between 1900 and 1920 the fossil fuels increased their combined share to 90 percent; they have maintained and even slightly increased that dominant position up to the present. However, whereas the bulk of the growth in energy consumption from 1880 to 1940 was fueled by coal, the liquid and gaseous fossil fuels have absorbed the brunt of the increase since then.

Nuclear energy, barely visible on the chart with one quarter of 1 percent of total energy consumption in 1969, is expected by many to provide the next major revolution in the composition of energy supply. Such a shift, however, is unlikely to be as rapid as either the displacement of fuel wood by coal or the displacement of coal by oil and natural gas. A major reason is that the contribution of nuclear

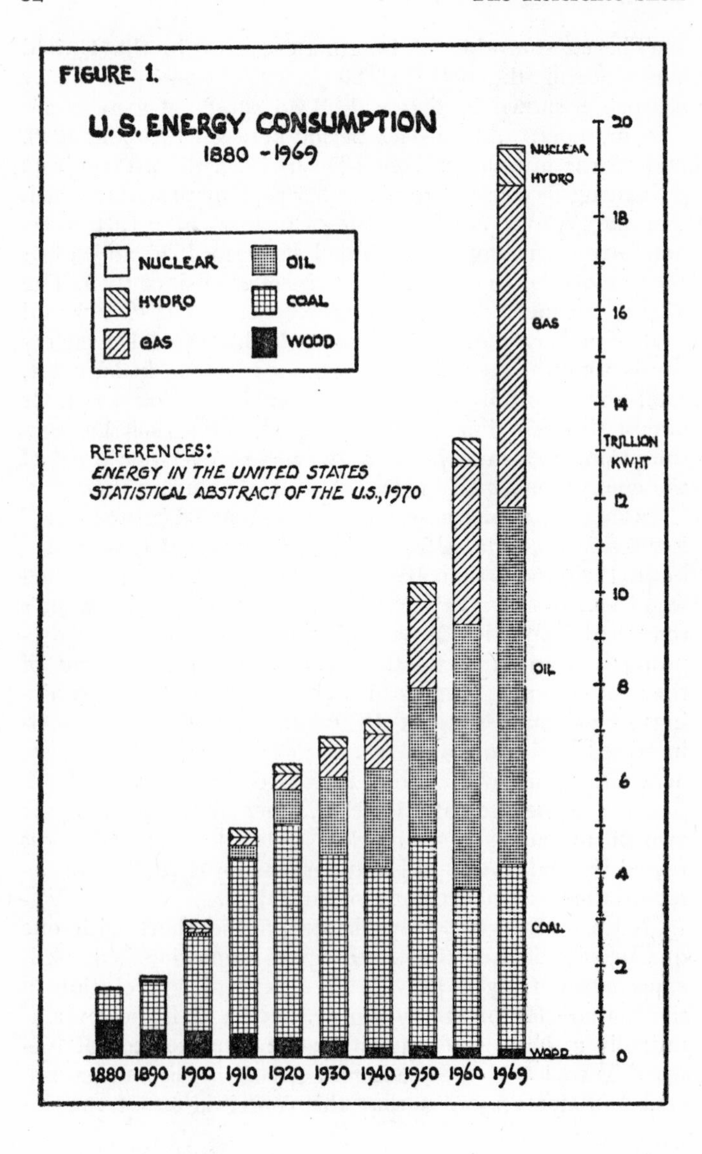
FIGURE 1.
U.S. ENERGY CONSUMPTION
1880 - 1969
NUCLEAR
HYDRO
GAS
OIL
COAL
WOOD
REFERENCES:
ENERGY IN THE UNITED STATES
STATISTICAL ABSTRACT OF THE U.S., 1970
NUCLEAR
HYDRO
GAS
OIL
COAL
WOOD
TRILLION
KWHT
20
18
16
14
12
10
8
6
4
2
0
1880
1890
1900
1910
1920
1930
1940
1950
1960
1969

energy today is restricted to the electrical component of the energy budget, and is likely to be so for some time. Nuclear reactors will probably remain impractical or uneconomical for most applications requiring a portable energy source. Their use as a source of raw heat for industrial processes does not seem imminent.

In recent years, electricity has received the lion's share of the public attention focused on energy problems, although it accounts for less than 25 percent of US energy consumption. As a "secondary" energy source, electricity is derived in all cases from one or the other of the "primary" sources, these being the fossil fuels, nuclear fuels, falling water, and—to a presently unimportant extent—tidal, geothermal, wind and direct solar energy. The use of electricity is increasing considerably faster than total energy consumption—since 1940 it has been doubling roughly every ten years—which accounts for the intense interest in electric power in general and nuclear power in particular. Even if the recent, dramatic growth rates were to persist, however, electricity would not account for 50 percent of total US energy use until about 2015. Nor is it really meaningful to consider electricity in isolation from other uses of the primary energy sources, because much of the growth of electricity consumption occurs when electricity is substituted for direct use of the fuels themselves.

Forecasts of Future Energy Consumption

A forecast is not the same thing as a prediction. A prediction means we think we know what will happen at some time in the future. A forecast is a much more cautious thing. It says that *if* a variety of conditions that hold today continue to hold in the future (including, for example, structure of demand and rates of growth), or *if* these conditions change in ways that are specified as part of the forecast, then such and such a future situation will be the result. People who make forecasts do not regard themselves as prophets, nor are they necessarily pleased with the prospects they are forecasting. In essence, they are telling us the probable consequences of present assumptions and present trends: if we do not like the consequences, we can work to change the assumptions and the trends. In this sense, many forecasters would like to be proven wrong. This certainly applies to the ecologists who have forecast worsening environment problems, and have been mistakenly called prophets of doom while working to avoid it.

Published forecasts of US energy consumption, based on growth rates observed in the early 1960s, range from 1.4 to 1.6 times the 1968 consumption in 1980 and from 2.5 to 3.3 times the 1968 figure in 2000 *(Man's Impact on the Global Environment).* If the growth rate of the late sixties persists, however, total US energy consumption will increase by 4.7 times between 1968 and 2000. Obviously, forecasts are very sensitive to the growth rate one assumes.

Forecasts for the electrical component of the energy budget as a separate entity are also available. A study prepared for the Joint Economic Committee of the United States Congress *(The Economy, Energy, and the Environment)* lists seven such forecasts for 1980 published since 1965, and one published since then for the year 2000. The forecasts for 1980 ranged from 1.8 times to 2.2 times the actual electricity consumption in 1968 (1.43 trillion kwhe). The single forecast for the year 2000 was 4.1 times the 1968 figure. Yet, if the average rate of growth of electrical consumption that actually prevailed between 1965 and 1969 persists, consumption will increase by 2.4 times between 1968 and 1980 and by 9.8 times between 1968 and 2000.

Finally, forecasts for world energy consumption deserve mention. Unfortunately, the data concerning what has happened up until the present are, on the whole, less complete and less reliable than those for the United States, thus making the already difficult art of forecasting even more so. Most observers agree that world energy consumption has been increasing at between 4 and 5 percent per year during the past few decades; forecasts for the next three decades assume rates as high as 5.7 percent per year. The latter figure leads to a doubling time of just over twelve years. Since this is faster growth than even the most sanguine promoter foresees for the United States, it would lead to a gradual reduction of the scandalous fraction of world energy consumption now accounted for by this country.

What Is the Energy Crisis?

The essential point about the foregoing forecasts for energy consumption is that they are not predictions and not inevitable. We can insure that they do not come to pass if we decide that the costs outweigh the benefits....

It will become apparent that although practically everyone agrees that there is an energy crisis, many people disagree about what it is. The energy industries have tended to regard the forecasts as inevitable and, indeed, desirable. They view the energy crisis as the problem of mobilizing technology and resources quickly enough to achieve the forecasted levels; to them, the growing opposition of environmentalists to their efforts is part of the crisis. (It should be said, though, that the word "crisis" rings hollow coming from utilities that continue to advertise in order to drum up demand.) The environmentalists agree about the existence of a crisis—they see it as the possibility that the forecasted levels of energy consumption might actually be achieved, accompanied by a level of environmental deterioration only hinted at today. And thoughtful observers on both sides worry about the costs of bringing pollution and depletion realistically into the balance sheets, and how the resulting increase in the cost of energy will affect the poor at home and abroad....

Two propositions, at least, seem relevant: first, as economists have known all along, a crisis in supply and demand can be met by moderating demand as well as by increasing supply; second, as biologists have known all along on this finite planet we must moderate demand eventually. It seems clear that, in terms of energy demand in the United States, the time for moderation is now, and that moderation is possible without returning to a primitive existence or disproportionately burdening the poor.

THE IMPORTANCE OF FOSSIL FUELS [4]

Energy needs today are met primarily by three fossil-fuel resources—coal, petroleum, and natural gas. Together they provide for over 95 percent of present demand, and by the year 2000 they will still meet over 70 percent of our needs. After fossil fuels, the other major sources of energy are hydropower and nuclear power. Hydropower now meets about 4 percent of our needs and nuclear power less than 1 percent. But by the year 2000 it is projected that hydropower will decline to little more than 3 percent, while nuclear power will supply over 26 percent of demand.

Coal is the nation's most abundant fossil-fuel resource, and it will continue to be an important contributor of energy to 1985 and beyond. Despite the abundance of coal, domestic use now meets only about 18 percent of our total energy demand, and its share is not projected to grow. Nevertheless, the rapid increase in energy consumption means that by 1985 the use of coal will rise by over 70 percent.

Environmental problems caused by the sulfur content of coal will restrict its use in many areas and will put pressure on the availability of low-sulfur coal. Development and implementation of technology to remove sulfur oxide emissions after combustion or to convert coal to a clean gas or liquid, however, are expected to extend use of coal while achieving air quality standards. There is also considerable concern about the adverse environmental effects of mining, particularly surface mining. The introduction of stringent mining controls should greatly reduce these detrimental

[4] From *Energy and the Environment—Electric Power,* a booklet prepared by the United States Council on Environmental Quality. '73. Supt. of Docs. Washington, D.C. 20402. p 5-9.

effects and thus enable continued production, but the time needed to install new productive capacity means that coal production may not be immediately responsive to rapid demand changes.

Oil is now the most intensely used energy resource—meeting about 44 percent of our needs. It will maintain this market share in 1985; in absolute terms, oil use will rise over 65 percent. And it will remain the largest source of energy through the end of the century. Unlike coal, however, demands for oil cannot be met solely from domestic production. Today we import 26 percent of our petroleum and petroleum products, and by 1985 we may import over half. Of our oil imports today, 40 percent is residual fuel oil. This increasing dependence on foreign sources of petroleum poses two major problems. First, greater reliance on potentially insecure sources will increase the risk to our national security and will adversely affect our balance of payments. In 1970, we imported $3.6 billion worth of oil and gas while exporting only $1.5 billion in energy fuels—primarily coal. By 1985, this balance of trade deficit for energy could rise to $25 billion annually. Although considerable oil reserves are believed to exist, the availability of oil depends upon many factors, including market prices, alternative energy sources, and the economics of resource recovery.

Natural gas now meets about 33 percent of our energy needs, but proven domestic reserves are limited. Unlike oil, however, gas imported from overseas is very costly. In 1971 we imported only 4 percent of our natural gas, and virtually all of that was piped from Canada. The limited availability of domestic gas supplies, as well as the high cost of importation, may severely limit the future use of natural gas, particularly if interstate gas sales are regulated at artificially low prices. By 1985 natural gas may meet less than 25 percent of all energy needs, although total gas use may rise by 34 percent. Virtually all of this increase would be from additional natural gas imports, including costly liquefied natural gas, and from synthetic gas production. In fact, by

1985, imports may supply almost 20 percent of our gas needs.

The limited availability of domestic natural gas supplies is of concern because importation raises national security and balance of payments issues analogous to those for oil and because gas is an environmentally desirable energy source. As with other fuels, there is a complex interrelationship of factors of national interest, including the environment, that determine the present and future availability of gas. Economic factors are of particular importance in discovery rates and in providing the equipment needed to exploit the proven reserves. For this reason, economic incentives can stimulate the discovery and production of the remaining sizable reserves of domestic natural gas.

These three fossil fuels, coal, oil, and gas, are the bases for many energy systems—for process heating, for transportation, and for electric power generation. Sixty-five percent of our coal is burned to produce electricity, while only 7 percent of our petroleum and 18 percent of our gas are used in power plants.

Unlike these fossil fuels, nuclear energy is used almost exclusively for electric power generation. At present, nuclear energy is only a small part of our total energy production, less than 1 percent. But by 1985 it may meet 10 percent of our total energy demand, rising to 26 percent of the total energy demand and 60 percent of electrical energy demand by the year 2000.

Water power too is used almost exclusively for the generation of electricity. Although historically it has been an important energy source, it currently meets only 4 percent of our total needs. And because there are few economical and environmentally suitable new sites for hydroelectric facilities, its growth potential is limited. By 1985, it may meet only 3.7 percent of our needs, and by the year 2000 the figure may decline to only 3 percent. Despite its declining relative importance, hydropower production is expected to more than double over the next thirty years.

Emerging systems requiring further technological development may eventually supply much of our energy, as liquid and gaseous fuels or as electricity, but their technical and economic feasibility has yet to be demonstrated. Should these new technologies be implemented, they can dramatically affect both our rate of use of particular resources and our ability to tap now uneconomic sources. Consequently, even short-term estimates of total energy use and energy system mix are subject to error, and longer-term projections are even more uncertain.

The overall demand for energy and the mix of energy systems have been strongly influenced by resource availability and, ultimately, by price. In the future, the relative price of energy systems will be influenced by technology, by resource availability, and to an unprecedented degree by a demand for environmental protection.

Whether energy resources are sufficiently plentiful to meet long-term demand is a question that is often raised. It would appear that availability in the physical sense is not a limiting factor. Coal resources are adequate for at least hundreds of years. But supplies of other fossil fuels under current price structures, particularly gas, appear in shorter supply. Petroleum, too, is in short supply, and we depend increasingly on foreign sources despite our considerable reserves on the North Slope of Alaska and in offshore areas. Even the supply of low-cost uranium, when used in nonbreeding light water and gas-cooled nuclear reactors, may be measurable only in decades. And coal, our most abundant fossil-fuel resource, may pose short-run availability problems as we seek to reduce sulfur oxide emissions and develop surface and underground mining requirements. Our concern about sulfur emissions has put pressure on low-sulfur coal reserves, and if we are to continue using high-sulfur coals, we will have to reduce the release of sulfur compounds to the atmosphere.

But these resource availability difficulties are not as ominous as they seem. Over the next decade, the use of control techniques to limit sulfur emissions could allow us to continue burning coal. Eventually, gasified and liquefied coal are likely to augment natural gas and petroleum supplies. Similarly, oil shale and other as yet uneconomic sources of fossil fuels may enter the energy mix. The use of breeder nuclear power plants could extend the lifetime of nuclear fuels almost indefinitely. Assessing the environmental effects of these alternative energy sources is, of course, critical to any decisions regarding their use.

Our short- and intermediate-term energy needs are of major concern. Because some energy resources are in short supply and because emerging systems are uncertain or require long start-up times, in the next decade we may have balance of payments difficulties as well as supply dislocations from domestic and potentially insecure foreign sources.

Another shorter-term energy issue is power plant construction. Generating capacity shortages (not energy resource shortages) have resulted from delays in construction of new power plants. These delays stem chiefly from construction, design, and operating problems. In a few cases, regulatory delays resulting from environmental conflicts have also been a factor. Enactment of the Administration's power plant siting bill would provide the institutional framework to avoid many of these difficulties.

The real question about the future is: How much are we willing to pay, in terms of resources, consumer prices, national security, and environmental quality, to satisfy our growing energy demand?

WHAT ENERGY CONSUMPTION DOES TO THE ENVIRONMENT[5]

Few realize the degree to which energy systems affect the environment, although many are becoming more aware of

[5] From *Energy and the Environment—Electric Power,* a booklet prepared by the United States Council on Environmental Quality. '73. Supt. of Docs. Washington, D.C. 20402. p 9-11.

damages from specific activities. Converting fossil and nuclear fuels into energy leads to air pollution, water pollution, creation of solid wastes, land disruption, and esthetic degradation. Automotive air pollution, thermal discharges to lakes and streams, and land destruction by surface mining typify the diverse environmental damages of energy consumption.

Air

Energy systems were the largest source of the 264 million tons of pollutants emitted into the air in 1970. Fifty-six percent of this total was carbon monoxide, with particulates composing 9 percent, sulfur oxides 13 percent, hydrocarbons 13 percent, and nitrogen oxides 9 percent. Automobiles and other forms of transportation caused over one-half the total emissions, contributing on a nationwide basis over 75 percent of the carbon monoxide and over 50 percent of the nitrogen oxides. Stationary sources, including power plants and residential and commercial heating units, were the second major component of total emissions, causing almost 80 percent of sulfur oxide emissions, and over 25 percent of particulate emissions. The absolute and relative importance of these emissions, of course, varies from one location to another.

Energy consumption also contributes environmental pollutants which can be highly toxic. In 1968, 14 percent of the lead utilized in the United States was emitted directly into the air from automobile tailpipes—a total of 180,000 tons. It is suspected that significant quantities of vanadium in the atmosphere result from residual fuel oil combustion, particularly from imported oils. Moreover, mercury escapes into the air when coal is burned. The effects on human health and the environment from these emissions are not yet well known.

Water

One consequence of oil exploitation is the discharge of oil into water. It can come from a drilling rig, pipeline, storage tank, or shipping accident, and on occasion the amount is massive. Oil spills can damage the aquatic environment, and if they occur near shore, they can destroy wetland spawning areas and prohibit recreational activity. Although accidents cause the large oil spills, the cumulative intentional discharges from oil extraction and transportation are still greater. They result from the incomplete separation of crude oil from the brines that are brought up with the oil and from the disposal of some brines in inland waterways and at sea. Discharge of these brines—several barrels of which are produced for every barrel of oil—can contaminate fresh water supplies and can affect marine ecosystems. Further, the cleaning of oil tanker bilges is a significant source of oily waste water discharged to the oceans.

Coal extraction is widely recognized as a source of water pollution. The Department of the Interior estimates that 13,000 miles of streams and 145,000 acres of lakes and reservoirs have been adversely affected by acid mine drainage and siltation from coal extraction.

The major source of thermal discharges into water bodies is from electric power plants. Of the 50 trillion gallons of cooling water used by US industry in 1964, almost 41 trillion gallons—81 percent—was for electric power plants. The effects of thermal discharges are not fully understood, but their raising ambient water temperature and altering the natural balance of aquatic life may degrade lakes, bays, estuaries, and rivers. The effects of these discharges can vary widely and in some cases may even be beneficial, depending on the rate and constancy of the thermal discharge, the size of the receiving water body, the climate, and the uses to which the water body is put.

Land

By 1965, surface mining had already affected over 3 million acres of land, of which only one third had been restored. At present mining rates, about 1 million additional acres has been affected since 1965. The surface mining of coal accounts for 41 percent of the total acreage disturbed. Unreclaimed, these lands are esthetically barren and preclude recreation, wildlife habitat, and other uses. However, in a few states adequate reclamation procedures are required; the introduction of national surface mining reclamation regulations could significantly reduce this aspect of environmental degradation. Deep coal mines have undermined over 7 million acres, of which over 2 million acres has already subsided. From now to the year 2000, about 4 million additional acres may be undermined from coal extraction. It may help in putting these figures in perspective to recall that Massachusetts has a land area of about 5 million acres.

Other energy activities also use large areas of land or cause esthetic blight. Electric utilities now operate 300,000 miles of overhead transmission lines involving 4 million acres of right of way. As electricity demand rises, they may require 4.5 million acres of additional land by the year 2000. Although the lines occupy narrow corridors of land, in total this projected land use is equal to an area larger than Delaware and Maryland combined.

Solid Waste

Energy systems generate solid wastes. Of the estimated total 4.3 billion tons produced in the United States annually, mining wastes account for almost 40 percent. Further, the processing and combustion of fuels, particularly coal, lead to significant amounts of solid wastes. In addition to occupying land, some wastes burn slowly, giving off noxious and dangerous gases. Runoff or leaching of some wastes impairs surface and ground water quality. Although present in large quantities, mining wastes are not highly toxic nor are they usually located near population centers.

Electric power generation from nuclear plants produces radioactive solid wastes that require careful disposal. About one million cubic feet of low-level solid radioactive wastes is now being generated each year, primarily by nuclear power plants, fuel fabrication plants, and reprocessing facilities. This volume, which is buried in federal- or state-controlled commercial burial grounds, is estimated to quadruple by 1980. In addition, the reprocessing of nuclear fuel elements produces high-level liquid radioactive wastes. The annual wastes from a 1,000-megawatt power plant, when solidified, would occupy approximately 100 cubic feet.

Federal regulations require that the spent fuel reprocessing plants solidify the high-level liquid wastes and ship them to a federal repository within ten years after processing the spent fuel. The site for such a federal repository has not been selected, however, and the AEC [Atomic Energy Commission] has proposed the use of surface facilities from which wastes could be retrieved for possible future transfer to a repository in a deep geological formation or for management in some other way. Because these wastes remain highly radioactive for thousands of years, continual monitoring and careful storage are of the utmost importance.

FEDERAL AGENCIES INVOLVED IN ENERGY [6]

More than 60 federal departments, agencies, commissions, and other bodies are engaged in activities which include specific concern with one or more aspects of the US energy supply. An estimated 44 agencies are engaged in the administration of specific energy-related programs, and the activities of an additional 20 have impact on the nation's energy situation.

[6] From "Controversy Over the Impact of Federal Regulation on the Petroleum Situation." *Congressional Digest*. 52:230-1. O. '73. Reprinted by permission.

The great majority of such federal instrumentalities are concerned in some degree with research, information gathering, or regulation of the petroleum and gas resources of the United States and with the individuals and organizations engaged in producing, processing, and distributing these energy sources. Among executive branch agencies with responsibilities affecting the gas and oil industries are the following:

Executive Office of the President

Within the Executive Office of the President several bodies have been constituted and assigned basic responsibility for coordinating governmental efforts and advising the President on matters of energy policy.

Energy Policy Office: Early in . . . [1973] President Nixon, in conjunction with one of a series of energy messages he has sent to the Congress, issued an executive order establishing an Energy Policy Office. As director of the office he appointed former Governor John Love of Colorado. . . . [He resigned in December.—Ed.]

In the order establishing the office, the President indicated that the director would serve as the President's principal energy adviser and would be responsible for identifying major problems, reviewing alternatives, making policy recommendations, assuring that agencies develop short- and long-range plans, and for monitoring the implementation of approved energy policies.

National Security Council: NSC advises the President with regard to policies relating to the national security, and reviews studies dealing with the ability of the United States to obtain necessary petroleum and gas required for any emergency.

Office of Science and Technology: OST performs a governmentwide function in the coordination of energy policy.

Council on Environmental Quality: The council coordinates federal activities as they relate to environmental quality, including those oil and gas matters that affect the environment.

Department of State

Office of Fuels and Energy: The office, a part of the Bureau of Economic Affairs, is responsible for the coordination of departmental policy and action in all foreign policy matters pertaining to petroleum.

Department of Defense

Many departmental functions are concerned with energy and fuels for the armed services, principally with matters such as fuel standards, procurement, stockpiling and storage, and research and development. Each major military service conducts similar activities. Of those direct military conducted operations in this regard which have broad impact on the nation's energy situation, one in particular should be noted:

Office of Naval Petroleum and Oil Shale Reserves: This office has the mission of exploring, prospecting, conserving, developing, using, operating, and administering the naval petroleum and oil shale reserves for the production of petroleum and shale oil when required for national defense.

Department of the Interior

Virtually all activities of the Department of Interior concerned with energy matters are the responsibility of an assistant secretary for energy and minerals. Among the major departmental functions involved are those carried out by the following Department of Interior organizational units:

Office of Oil and Gas: This office was established to serve as a focal point for leadership and information on petroleum matters in the federal government, and to be a channel of communication between the federal government and the petroleum industry as well as the oil-producing states. Among other functions, the office allocates and licenses petroleum imports to qualified applicants.

Bureau of Mines: The bureau conducts investigations and performs research on petroleum and natural gas, including basic research on oil shale. Additionally, it produces helium from natural gas and conducts research on synthetic fuels.

Office of Energy Conservation: The office was created in 1973 to promote efficiency in the use and development of energy resources, to coordinate all federal energy conservation programs, to conduct research on methods of improving the efficiency of energy usage, to promote consumer awareness of the need for energy conservation, and to develop contingency plans for nationwide power, fuel, and mineral resources emergencies caused by natural disasters or other interruptions of the nation's energy and mineral supplies.

Geological Survey: Collects, analyzes and distributes technical geological and geophysical information and conducts research on the mineral resources of the nation, including petroleum and gas. It classifies the public lands as mineral or nonmineral in character and supervises the operations of oil and gas leases on public lands, Indian lands, certain acquired lands, and the continental shelf.

Bureau of Land Management: Issues and administers oil, gas, and oil shale mineral leases on the public domain, acquired lands, and submerged lands. The bureau also processes requests for rights of way of petroleum and gas pipelines over the public domain and the outer continental shelf. Conducts supply and demand analyses to determine the location, size, and timing of offshore oil and gas sales.

Bureau of Indian Affairs: The bureau administers oil and gas leasing matters, and the income derived from these sources, on Indian land held in restricted and trust status by the United States.

Office of Hearings and Appeals: Included in the office is the Oil Import Appeals Board, consisting of representatives from the departments of Interior, Justice, and Commerce. It considers appeals and petitions by persons affected by oil import regulations.

In addition to the above functions performed within the department, the Secretary of Interior serves as chairman of the federal government's Oil Policy Committee, whose members also include the secretaries of Commerce and the Treasury, or their designees.

Department of Commerce

Maritime Administration: Is responsible for the development and maintenance of a US flag merchant fleet, including tankers, adequate to meet national defense and commercial requirements.

National Oceanic and Atmospheric Administration: NOAA is responsible for determination of seaward boundaries with reference to the coastline, important to the petroleum industry's plans concerning future oil and gas matters in offshore areas.

Department of Transportation

National Transportation Safety Board: The board may, on its own motion, conduct investigations of pipeline accidents or rail, highway, or aviation accidents involving the transportation of petroleum or natural gas. It may also participate in such investigations undertaken by the Department of Transportation. It may also undertake special studies involving pipeline safety or the transportation of petroleum or natural gas by other modes.

US Coast Guard: The Coast Guard enforces the Oil Pollution Act of 1961.

Federal Power Commission

The commission is charged with the administration and enforcement of the Natural Gas Act. Its activities are of a regulatory nature over the transportation and sale for resale of natural gas in interstate commerce, including the establishment of wholesale rates, the granting or denial of authority to construct, acquire, operate and abandon natural gas facilities, initiate or discontinue natural gas service, and to import and export natural gas to and from the United States subject to favorable recommendations of the secretaries of the departments of State and Defense.

Interstate Commerce Commission

The ICC regulates common carriers, including oil pipelines, that are subject to the Interstate Commerce Act.

Environmental Protection Agency

National Air Pollution Control Administration: Utilizes information on fuels obtained from government agencies and industry with respect to availability of fuels, their quality and demand, impact of fuel consumption on air quality, . . . to guide it in research and development programs.

Federal Water Quality Control Administration: This agency is responsible for protecting and enhancing inland and coastal waters of the United States. Whenever oil is or threatens to be discharged to water, the FWQCA takes action to prevent or mitigate damage to the environment and recommends enforcement action against violators.

Cost of Living Council

This agency establishes and oversees enforcement of price levels for certain commodities and consumer and industrial supplies and products, including gasoline and other petroleum byproducts.

II. THE FOSSIL FUELS: COAL AND SHALE

EDITOR'S INTRODUCTION

There is much speculation about power from windmills and tides, and surely in the future these "free" energy sources will be utilized. Fission and fusion and fuel cells are all possibilities as well. Doubtless technicians will find ways to make use of at least some of these fuel sources.

But the plain fact is that right now the overwhelming preponderance of energy actually used in the United States comes from the old-fashioned fossil fuels. They are familiar standbys: coal, oil, natural gas. Once used, they are gone. They do not renew themselves. They were created in the earth over millions of years, and we are consuming them in decades.

The exact figures vary according to the authorities cited, but 1973 federal figures indicate that 95 percent of present energy demand is met by the three primary fossil fuels. Coal today lags behind oil and natural gas because it is dirty, gives off sulfur when burned, and presents major environmental hazards. Thus 75 percent of present US energy needs are supplied by oil and natural gas.

This section on fossil fuels begins with two articles that explore the problems and possibilities of increasing the use of coal. There is more coal in the United States than any other energy source now available. The supply is estimated to be enough to last anywhere from three hundred to eight hundred years at current rates of consumption, but such projections assume that the United States will be willing to pay a high price in environmental damage by mining. An article by Jane Stein, "Coal Is Cheap, Hated, Abundant, Filthy, Needed," opens the discussion and includes an explanation of coal gasification—a way of turning coal into a synthetic gas and therefore a "cleaner" fuel. The next article describes a conflict in Montana between a coal company that plans to strip-mine ranch lands, and the ranchers, who worry about water tables and whether the land can in fact be replaced in such a way that the grass will again grow.

The final article in this section discusses shale, a rock that contains oil and is found in several western states. There is a great deal of shale in the West, much of it on government-owned land. So far no one has solved the financial problems of extracting the oil from the rock economically or the environmental problems of what to do with the vast amounts of slag that result from the extraction. As the energy squeeze gets worse, however, efforts to extract oil from the rock will increase.

COAL IS CHEAP, HATED, ABUNDANT, FILTHY, NEEDED [1]

King Coal is dead. Long live King Coal.

After decades of declining production and increasing disfavor, coal, our most abundant energy resource, has been staging a comeback. It is one of the ironies—and dilemmas—of our environmentally aware age that we will use more, not less, of this filthiest of fuels.

Underground coal mines have polluted the water table, harbored fires, some of which have been smoldering for decades, and caused millions of acres of surface land to subside, breaking roads and sewers and collapsing houses. Coal miners are often trapped underground; many more contract the miners' disabling disease, black lung. Strip mining, although twice as safe, has left lunarlike landscapes in its wake and polluted rivers and water supplies with silt and acid drainage. The blight of coal continues into the burning process, befouling the air with 60 percent of the fourteen million tons of sulfur dioxide discharged a year by US smokestacks.

[1] From article by Jane Stein, writer on energy and thermal pollution. *Smithsonian.* 3:18-27. F. '73. Copyright 1973 Smithsonian Institution. Reprinted by permission from February 1973 *Smithsonian* Magazine.

Yet the energy demands of a growing industrial society are doubling every ten years while the nation's energy resources are dwindling—and this is the crux of the energy crisis that has suddenly forced its way into our consciousness. National reserves of natural gas—the only widely used fuel that does not add pollutants into the environment—have dwindled, and demand has outstripped the supply. Domestic production is expected to peak in the mid-seventies and decline thereafter. Old gas customers across the country are rationed, new ones rejected, and the United States is being forced into heavier dependence on foreign suppliers—including Algeria and perhaps the Soviet Union—resulting in far more expensive gas.

Oil is in short supply—current oil reserves in the contiguous forty-eight states are now at the lowest point in twenty years—and we rely more and more on high-priced imports from the unstable Middle East. Nuclear power accounts for less than 2 percent of the nation's energy production and, owing to a series of setbacks including construction delays and legal proceedings initiated by environmental groups over reactor safety, it is far from fulfilling its promises on schedule. The nuclear breeder program is due on the scene no earlier than 1985, and even this program is beginning to find serious new critics. Pushed on the back burner by low funding are potential clean sources of electricity, such as solar energy and fusion. . . . Energy conservation, perhaps the most fruitful avenue for averting an immediate crisis, is just beginning to be discussed.

In desperation over coming energy shortages, energy planners are turning again to dirty but ever abundant coal, found in thirty-eight states with some one and a half trillion tons of it still under American soil—more than one thousand years' worth at today's recovery and consumption levels. The new wave of coal promotion is well under way already.

Coal is fueling one of the biggest as well as one of the most disputed electric power projects in the country—the

Four Corners power plant at Fruitland, New Mexico, near the joint state borders of New Mexico, Arizona, Colorado and Utah.

The Navajo strip mine that supplies coal for the Four Corners plant is one of the largest producers of coal in the country; some 22,000 tons daily are gouged from the reservation. Reclamation efforts so far have been failures, and the big generators spew some 350 tons of fly ash daily into what used to be the purest air in the country. Indeed, so bad has the haze quickly become from the Four Corners complex that its very smoke plumes are not visible beyond one hundred miles. And all this is to provide electric power for places as far away as Los Angeles—where air quality standards forbid the spewing of these pollutants.

Four Corners is the first of a planned coal-fueled system of at least half a dozen plant complexes in the Southwest, including another already completed plant complex at the southern tip of Nevada. The coal for this plant, ripped from a sacred Hopi mountain, is ground up at the mine, mixed with the region's precious water and piped more than 270 miles through a slurry line to the plant.

A Silk Purse in the West

This is only a taste of what may be coming. A government geologist sees the United States on the brink of "generations of strip-mining for coal that will make the excavation for the Panama Canal look like a furrow in my backyard vegetable garden." Twenty-year forecasts call for 300 million tons of coal a year—about half of our yearly production—being processed by huge, refinerylike plants, surrounded by massive strip mines in the western coal fields. But the product won't be coal—it will be gas, quadrillions of cubic feet of pipeline quality, pollution-free *gas*. We are promised a silk purse from a sow's ear.

Gas from coal—the process is called coal gasification—is a complex chemical transformation that changes nuggets of dirty coal into gas much like natural gas. What happens is that coal is changed into gas by adding hydrogen. To get the extra hydrogen, coal technologists turn to water.

Simply stated, the process is as follows. Water is heated to produce steam which reacts with the carbon from coal to make a hydrogen-rich gas. In addition to hydrogen, other gases such as carbon dioxide, hydrogen sulfide and ammonia are produced. The next step is to "clean" the gas, to purify it and separate out the unwanted constituents, leaving only carbon monoxide, hydrogen and methane. What is left is combustible but low in heat content compared to natural gas.

The final step—the hardest one and one that is yet to be proved in a commercial synthetic plant—is to upgrade the gas in heat content and cleanliness. This process, called methanation, calls for further reactions of the carbon monoxide and hydrogen to produce yet more methane (which is what natural gas is). All of this takes place at intense temperatures (upwards of 1,500 degrees Fahrenheit) and at pressures of more than 1,000 pounds per square inch.

The technique is an old one and goes back to the late eighteenth century when it was first used for lighting. Coal gas flourished until the advent of electric lighting and was subsequently used for heating. But then natural gas was turned from a useless and wasted byproduct of oil (wells that gave off gas were flared and allowed to burn themselves out) into the cleanest, cheapest source of energy. In the 1920s and 1930s natural gas became available by pipeline, and the market for manufactured gas gradually evaporated.

The gas derived from coal a century ago was essentially a byproduct; the primary product was coke, which was then sold for home heating and steel furnaces. Since the market for coke for home heating no longer exists—imagine our air pollution problems if coal was still used extensively for heating—the old techniques for coal gasification are impractical. In addition, "town" gas was low in energy content by today's standards. The new coal gasification technology aims at getting a clean-burning gas at the same heat content as natural gas—1,000 British Thermal Units (BTU) of heat per cubic foot.

Only one commercially proved coal gasification method operates today, and at that it produces gas at less than half the heating value of pipeline gas. This is the Lurgi process, developed in Germany during World War II, and still at work at plants in Western Europe, South Africa and Australia. Lurgi will make its American debut in New Mexico at two different sites in the mid-to-late 1970s. In addition, several gas companies are negotiating with Lurgi to build plants.

The Lurgi process produces a low-quality gas (400 BTU) which can be blended into the main transmission system of the 1,000-BTU gas currently being transported. But blending low-energy gas into high-energy gas has obviously limited possibilities, and engineers hope to devise a system to add to the Lurgi system which will raise the BTU value. A drawback to the Lurgi process is that it can accept only lump coal as opposed to the coal fines—dust and small coal particles—that are the product of strip mining. The New Mexico sites will provide the needed lump coal, but most of the coal fields now being eyed in the West will have to be strip-mined.

This is just one of more than a half-dozen coal gasification techniques under way at various stages of development, ranging from some in design stages through existing pilot plants. There is no single agreed-upon technology for coal gasification. At each step in the process there are several possible ways of designing the system, and most energy planners in the federal government are willing to support various approaches for now, believing that they cannot afford to decide prematurely on the wrong process.

Speeding Up the Pace

But energy planners looking at the near future hope to speed up the pace. Through most of the sixties, industry was not pursuing coal gasification and the Department of the

Interior's Office of Coal Research [OCR] had a low priority. "We were swimming against the tide," said George Fumich Jr., OCR director, pointing out that some of their projects have been ongoing for nearly ten years.

It is only recently that coal gasification has found itself on center stage. It received its biggest boost from President Nixon's June 1971 energy message in which he called for a "greater focus and urgency" to developing the needed technology.

Only last December [1972], New York State's Public Service Commission issued a report warning that nuclear power was not the answer to all the state's problems, that the breeder reactor was not only far off but equally fraught with problems. The commission called for the rapid development of fuel cells and, significantly, of coal gasification.

In August 1971, after the President's energy message, an eight-year agreement was signed between the Department of the Interior and the American Gas Association to develop joint facilities. An estimated $300 million will be spent with the hope of building a small demonstration plant—about one-third capacity of a full-scale operation—in the middle or late 1970s. To bring it up to commercial size, producing 250 million cubic feet of gas a day, two or more operating units would be added in 1980. (So far, engineers are working on pilot plants that, they hope, will produce a mere 1.5 million cubic feet of gas a day when on full stream . . . [in 1937].)

Officials at [the Department of the] Interior are confident that commercial facilities are just ten to fifteen years away. Former Assistant Secretary Hollis M. Dole predicted . . . [in 1972] before his resignation, that coal gasification will produce 2 percent of the nation's gas requirements by 1985—which obviously won't make a great impact on our energy consumption patterns. But by 2000, Dole anticipated, coal gas would account for as much as 10 percent of the total gas demand. The rest of the gas supplies would come from US natural gas reserves, liquefied natural gas piped in from Alaska or Canada, or tanked in from as close as Trinidad and as far away as Siberia. Boeing has circulated an artist's conception of a huge airborne tanker designed to haul liquefied natural gas vast distances. Additional supplies may come from synthetic natural gases made of naphtha and crude oil—techniques which are now being developed.

Coal gasification's value as an energy source is twofold. In the relatively short term it can help alleviate a tight fuel situation; in the long term, it can be used as a supplemental resource and, more importantly, as a "shelf" item—a dependable source that is always available, one that can be ordered and delivered within a few years.

Coal gas will be expensive—at least $1 per thousand cubic feet, about double that of natural gas today. But, assuming a severely declining natural gas supply, Fumich reasons that "pipeline gas from coal will be just as cheap—if not cheaper—than any other supplemental source such as gas from Algeria or Alaska or Russia. These imports are our gas alternatives and they are in the dollar and up bracket." Indeed, many critics have long held that natural gas prices have been artificially kept excessively low. They will inevitably have to rise now, but gas may not lose all its competitive edge since the prices of other fuels will surely rise also.

Coal gas as a *fuel* gets good marks from the environmentalists: It is a way of using coal, which is becoming difficult to burn under the ever-increasingly strict air quality standards. (In recent years, New York City, Boston and New Jersey, among others, have issued sulfur standards that virtually ban the use of coal as a fuel.) Coal gas creates little noise, water or air pollution, and requires no unsightly transmission lines since it will use already existing pipelines. But there are very serious environmental problems elsewhere in the coal-gas *system*. Coal is inescapably a dirty product: Getting it from the ground and converting it to gas are inherently dirty processes.

Coal contains up to 5 percent sulfur and 20 percent ash, both of which are harmful to the environment if spewed into the air, falling down as sulfuric acid rain, defoliating trees and bushes on the way and leaving a film of fly ash.

Sulfur must be removed in the coal gasification system because it acts as a poison to the catalyst. One solution is to collect the sulfur wastes through known and commercially available methods such as scrubbers, then collect and sell the waste product. But as a result of other antipollution efforts, sulfur is so overabundant that some natural sulfur mines may have to be shut down.

Ash can be used in the construction of cinder blocks or, if the transportation problem could be solved, it could be used for filling in strip-mined areas. While nothing would grow on it, ash used as landfill could be covered with good soil upon which vegetation would grow. Incredibly, some of the pilot projects currently under way have made no provisions for handling these wastes.

Water use is another potential problem: Vast quantities are needed for the hydrogenation process and for cooling purposes. Estimates are that a fully water-cooled coal gasification plant capable of producing 250 million cubic feet of gas a day would consume about 30 million gallons of water per day. That is only half of the water needed for a conventional steam power plant, according to Interior [Department] estimates, so the problem isn't how much water—but where all the water will come from.

Water consumption in the West—where most of the potential coal gasification sites are—has long been a problem. The big river basins, the obvious sources of water, are already overtaxed. So great is the water problem that future gasification sites will probably be limited by water requirements, not coal reserves.

The major objection, however, to coal gasification development remains the mining of the coal. The key is low-cost coal and this means strip-mined coal—with the specter of Appalachian horrors spread across the country. (One alternative—still in the experimental stages—is to convert to gas underground, called *in situ* gasification. It involves breaking up the coal formations with injections of chemicals and air, which then react with the coal to form gas. But nothing is problem-free: The environmental hang-up here is gradual land subsidence as the coal is converted to gas.) Strip mining remains the most alluring method because output per man-day is roughly double that of deep-pit mining, and operating costs are much lower. In 1970, strip-mined coal sold for $4.69 a ton; deep-mined coal was $7.40 a ton.

The Problem of Reclamation

Currently about 44 percent of all coal mined comes from stripping—a mere trickle compared to what is ahead. There are huge reserves of strippable coal in the West, particularly in Wyoming, Montana, New Mexico, North Dakota and Texas. This is mostly low-sulfur coal (which is good), but it also produces especially high amounts of fly ash (which is bad). Coal, petroleum and pipeline interests, along with land brokers and speculators, are already scrambling to assemble leases and rights to large tracts in these areas. And in the East—already spotted with "wasteland" areas—gas companies are busy seeking out options on major coal reserves in Appalachia, wherever there are sites large enough to support gasification plants.

The coal boom is bringing deep concern in its wake. Among the worriers is Senator Lee Metcalf (Democrat, Montana): "The fossil-fuel development now contemplated in Montana and neighboring states will affect the land, the quantity and quality of water available for other uses and the air. If we don't have immediate and effective federal and state legislation to control strip mining, eastern Montana will become a wasteland." The government and the gas and coal industries play down these concerns about strip mining. "There is no reason," says Douglas King, director of research and engineering at the American Gas Association, "why strip mining has to be so bad. The land can be reclaimed, but it is a question of economics."

Indeed, reclamation is often, though not always, possible —but among coal developers it is unpopular. Costs run as high as $2,700 an acre, and strippers rarely spend more than $300 on rehabilitation—if they spend anything at all.

Meaningful reclamation of strip mines would add to the cost—anywhere from 10 cents to 50 cents per ton of coal. The average consumer would pay about 15 cents a month more for electric power, calculates Representative Wayne L. Hays (Democrat, Ohio), who is a supporter of tough anti-stripping legislation. The public is willing to pay this small amount extra to protect the land from devastation, he declares.

Of the more than 1.5 million acres of American land stripped for coal, two thirds are unreclaimed, the Conservation Foundation estimates, and the unreclaimed areas keep on producing acid drainage, erosion and esthetic blight. Even where there has been reclamation, it has met with mixed results. In flat areas, some reclamation efforts have been successful, but they seldom work in steep mountainous terrain. The most popular method of reclaiming is simply to level the top of the spoil site, rather than regrading it to the original contour.

There is a particular danger in strip mining in arid regions where water supplies are overtaxed. In many such areas, the natural vegetation can and will return eventually if—and it is a big if—the stripping process has not lowered the water table appreciably. Reclamation is *not* just a matter of economics.

While there is no total ban on strip mining (some states have partial bans), Representative Ken Hechler (Democrat, West Virginia) has been an outspoken advocate of legislation to prohibit it altogether, a plan that seems unlikely to achieve much congressional support. (West Virginia's Democratic gubernatorial candidate, Jay Rockefeller, took a stand against strip mining and lost in the 1972 campaign.) John McCormick of the National COALition Against Strip Mining, a lobby group representing more than twenty environmental groups, feels that if coal means so much to us as an energy source (as it should), and if stripping is the only way to extract it (as it is in much of the West), then "we should see about getting the best reclamation possible without worrying about costs. Plant roses if that's what will produce the best results."

Enforcement of tough reclamation regulations is clearly the major issue for environmentalists. State laws have provided exceedingly meager protection. More than half of the states have statutes requiring at least some degree of land restoration, but they are often undercut by ineffective backup regulations and limp enforcement.

At the federal level, there is no systematic strip mining legislation. The only federal control consists of Interior Department regulations for mining on public and Indian lands over which the government has jurisdiction. Even here, inspection capabilities are inadequate, and there are no provisions for public participation. Nor do even the limited federal regulations apply retroactively: The vast acreage of strippable coal leased before the regulations went into effect in 1969—some 2.4 million acres—are beyond the law. . . .

Without stringent reclamation laws, coal gasification will be little more than a tradeoff of one pollution for another. It would help clean up urban areas with a cleaner fuel, but it would blight vast rural areas that are now relatively free of pollution and environmental degradation.

The second reign of King Coal is not likely to be a tranquil one.

ENERGY AND FOOD NEEDS CLASH IN WESTERN STATES [2]

A cavalcade of tour buses rumbled through the semi-arid sagebrush country . . . near Colstrip [Montana], . . . the center of one of the most extensive strip mining developments in the northern plains.

Behind lay the gouged open black pits in which gargantuan mechanized draglines were scooping out coal. Ahead lay an untouched pasture.

An energy company representative announced over one bus loudspeaker that this pasture, too, would soon be dug up. "How soon you gonna strip it?" asked a man in a Stetson in the back of the bus, " 'cause I farm it now."

The dialogue epitomized a struggle with international implications that has been joined, not only in Montana but also in all the coal-rich western states between the forces of agriculture and energy. . . .

Put in the simplest terms, states such as Montana, Wyoming, North Dakota, Colorado and New Mexico find themselves caught between the federal government's policy of utilizing more home-grown fuel to offset dependence on Middle East oil, and the demand of their local farmers and ranchers, under whose land lie billions of tons of coal.

Opposing Positions

The energy companies say coal development and agriculture can develop side by side. Farming, ranching and environmental groups insist they cannot, because the two industries compete, they say, for water, land, clean air and labor. Pointing to the world food shortage, one local group, the Rosebud Protective Association, stated its case by declaring, "It is senseless for the United States to unnecessarily sacrifice food-raising potential to energy development."

[2] From article by Grace Lichtenstein, staff reporter. New York *Times*. p 12. Ap. 2, '75. © 1975 by The New York Times Company. Reprinted by permission.

While state officials prepared to debate the issue at the Governors' Conference, more than a dozen Rosebud cowboys gave them an on-site preview by inviting themselves on the Colstrip bus tour, which was sponsored by several energy companies.

"I'm just spreading heat and discontent everywhere I go," said Wallace McRae, a local activist rancher with the face of a Marlborough man and the politics of a western Ralph Nader, with only a trace of a smile. "To have the company conduct this tour without showing the agricultural side is a travesty."

Colstrip is the focus of the energy-agriculture battle in Montana. It is a desolate boom town filled with mud, trailers and some new rental houses that look like trailers with sloped roofs, surrounded by stripped coal ditches and dominated by two 500-feet gray concrete smokestacks. They are part of two 350-megawatt electric power plants now being built to burn the coal straight from the neighboring mines. More of the resulting electricity would go to states in the Pacific Northwest than to Montana.

Many ranchers are resigned to the current operation. But they hopped on the buses to argue against further encroachment of their land, to snicker at energy company claims about returning stripped land to productive agricultural use and to lobby against the proposed construction of two additional giant generating plants at Colstrip.

Representatives of the Montana Power Company handed out slick four-color brochures describing Colstrip as "an exciting combination of pioneer spirit and a modern, well-planned community."

"Balderdash," said Wallace McRae.

The company said that in six years of mining, its coal operation had disturbed only about five hundred acres of land, with all but twenty-five acres under some phase of reclamation.

"But what they don't tell you is that's the heart of the land, the very best soil. It's like taking the heart out of a man. It doesn't look like much, but he's no good without it," said Nick Golder, whose ranch is sixteen miles from Colstrip.

Mr. Golder belittled the reclamation areas—patches of land that had been recontoured after the coal was taken out, heavily fertilized and then replanted with a variety of grasses. "I know grass," Mr. Golder said, "it's my business. This looks terrible now. What's it going to look like in fifteen years?"

"I don't know. I won't be here in fifteen years," replied a power company spokesman.

More than anything else, the ranchers worry about their water in a brown plains area that gets only fourteen inches of rain on average a year. They irrigate their land with wells that draw from aquifers—underground springs. There is evidence that strip mining dams up some aquifers.

"Duke McRae lost two springs recently," said Nick Golder. "These lands are useless for grazing without the water."

The state government worries about the whole concept of Colstrip. In 1973 it passed one of the nation's toughest laws dealing with the sites of power plants. Under that law a hearing is set to begin in three weeks to determine whether the energy companies should be allowed to build their third and fourth giant plants at Colstrip.

The Montana Department of Natural Resources, after an exhaustive study, came out against the new plants, saying they would create air pollution, water trouble and social turmoil. The decision on the extra plants "will set a precedent for the rest of Montana and possibly the nation," a spokesman for the Natural Resources department said.

SHALE OIL: TANTALIZING, FRUSTRATING [3]

Take Highway 70 east out of Grand Junction past the town's famous peach orchards, its new, hopefully planted

[3] From article in *Forbes*. 115:57-60. Ap. 1, '75. Reprinted by permission of Forbes Magazine from the April 1, 1975 issue.

grapevines, its picturesque buffalo herd. Wind with the salty Colorado River between flat-topped mountains, in and out of swirling snowstorms, then bright sunshine. Arrive an hour or two later in Rifle, Colorado, one of the hunting capitals of the world. More than this, however, Rifle is a place where townspeople have grown up knowing that all around them, especially to the northwest, is a vast amount of oil—1.8 trillion barrels at latest estimate, more than in the entire Middle East. If even 1 percent or 2 percent of it could be brought to market, the United States' energy gap would narrow. But to everyone's frustration, the oil is locked inside rocks and most of the rocks buried under several hundred feet of earth.

F. M. (Mike) Cross is the local jeweler. Years ago he began polishing the rock he picked up in his own backyard, cutting it into pieces and making jewelry. Today Cross sells buckets of his "Kero-gems" not just to summer tourists but to the oil companies prowling the landscape. (Kero is for kerogen, the name of the black stuff in the rock that turns to oil on heating.)

Cross may turn out to be the only man ever to make real money from shale oil.

That's not to say others aren't trying. Ever since men first discovered that those outcroppings were soaked with oil, they've been trying to figure ways to get at it. It may be just one of those tall western tales, but the locals like to tell about the sheep rancher who invited his neighbors to his housewarming, built a fire in his "stone" fireplace, and saw the whole thing go up in a stinking blaze.

Since then there have been a whole series of shale oil booms, coinciding, for the most part, with the energy crises that preceded most oil discoveries in this century.

In every boom, men dreamed and schemed of efficient ways to (1) mine the rock, (2) heat it to 900° Fahrenheit, (3) collect and condense the vapor—oil—that is released, (4) recover any other valuable elements present and (5) dispose of the remaining rock which, by a perverse law of nature, expands considerably after the oil is removed.

For the dreamers and schemers, hopes, long dormant, soared again after the Organization of Petroleum Exporting Countries [OPEC] applied the screws to the oil users of the world. The federal government, which owns 80 percent of the 11 million acres that is shale country, moved—putting up a wee bit of its land for competitive bidding. One outfit, Colony Development Corporation, announced plans (later canceled) to build a 50,000-barrel-per-day commercial oil shale plant starting . . . [in the spring of 1975].

Now, however, it's not at all certain that even $10-per-barrel oil can bring shale to market. The biggest problem is cost: With known processes, estimates are that it will require $12,000 in capital for every barrel of commercial production—and inflation is raising the figure almost daily. This compares with $8,000 per daily barrels for North Sea oil and $5,000 to $10,000 for Alaskan.

Another horrendous problem is water, which is required in most processes. By a bitter twist, the shale is located in some of the driest areas in the United States. The primary water supply is the Colorado. Every drop taken from it increases its salinity, which has serious implications for agriculture and for US relations with Mexico.

And finally, what do you do with debris—after you have mined out the rock and removed the oil?

Field of Battle

So far, none of these problems has been solved satisfactorily. And yet the quest continues, quickened by the high cost of oil and the United States' growing vulnerability to imports. It is centered in one area—bounded by Denver on the east, Grand Junction on the west, Vernal, Utah and Laramie, Wyoming on the north.

"We're going to do it this time," John B. Jones Jr. insists. Jones has made shale his life's work. A chemical engineer, Jones says: "Before World War II, I worked for the Bureau of Mines in Bartlesville, Oklahoma. My boss had been involved in oil shale in the twenties. That was my first exposure to it." Later, Jones started a consulting business and got involved in Brazil, helping the government there set up an oil shale operation that is still producing oil.

In 1966 Standard Oil of Ohio [Sohio], in the person of Harry Pforzheimer, also a chemical engineer by training, discovered Jones. Sohio, crude-deficient and eager to bolster its domestic reserves, was fascinated by shale. It became one of the original partners in Colony Development Corporation, which first demonstrated its Tosco II process in 1964.

The Tosco process, owned by Oil Shale Corporation . . . works. Unfortunately, it is complicated, expensive and water-intensive. It first heats ceramic balls, then uses the balls to heat the previously mined shale in a rotating vessel. It requires considerable quantities of water because it reduces the rock to dust, which must be damped down. So, Sohio looked the world over for a simpler, less expensive process.

The process they chose was Jones's. But since Sohio had to raise several billion dollars for the Alaska pipeline, Pforzheimer put together a separate group to fund the new oil shale process. "Our plan," he says, "was to go to the public, but investment bankers advised against it, so we raised $500,000 from each of seventeen companies." Sohio is one of the seventeen.

The new company, Paraho Oil Shale Demonstration Inc., leased the old Bureau of Mines site at Anvil Points and got started. Six months later Jones had a pilot plant operating. Now the outfit has just completed a 45-day run of a larger "semiworks" plant, and both Pforzheimer and Jones are grinning from ear to ear. In Jones's process, the shale is fed into a stationary retort; it moves through a fire zone where the oil comes off as a vapor and is removed. The spent shale flows out the bottom.

"We've *got* it this time," Jones repeats softly.

Maybe so, but the commercial-sized test is yet to come. That is planned for later . . . [in 1975].

Several canyons to the west of Anvil Point lies Logan Wash, just outside De Beque, which was the center of the 1920s oil shale boom. Here another group of engineers is looking very pleased with themselves. These men work for two subsidiaries of Dr. Armand Hammer's Occidental Petroleum Corporation—Garrett Research & Development Company and the newly formed Occidental Oil Shale Inc. [Oxy].

"This is the right field and the right process," claims Richard Ridley, chemical engineer and manager of Oxy's shale project.

Occidental's process heats the shale while it is still in the ground. This has great appeal because it essentially eliminates two very expensive and environmentally controversial steps: mining the rock in the first place and then disposing of it once the oil has been removed. It does not require vast quantities of water.

Dr. Donald E. Garrett, head of Garrett Research, says he came up with the basic idea after Hammer had sent him off on two different occasions to look over the Colony operation with a view toward buying in. Garrett advised "no" twice, he relates, and the irrepressible Hammer essentially said: "If you're so smart, come up with your own process."

Garrett's basic idea was to mine out a relatively small "room" under the oil shale. From there, workers would drill holes up into the shale for explosives to fracture it, and other holes down into it for oxygen, for starting a fire and for heat-measuring devices.

The idea is for the heat to release the oil, which would flow to the bottom to be drained off. Some of the gases would be removed for other uses and some would be recirculated to keep the fire burning. But there are problems—such as how to control the explosions, how to keep the fire burning evenly, how to drill through rubble—which are keeping Oxy's engineers busy.

From the beginning Oxy's whole concept has been greeted with considerable skepticism—partly because of

Hammer's reputation for exaggerating the company's accomplishments and partly because Oxy has insisted on keeping a high degree of secrecy up in Logan Wash. . . .

The Hopefuls

Union Oil of California is also in the act, has been since the late forties. Not much is going on at the Union Oil site on Parachute Creek halfway between Logan Wash and Anvil Points right now—there isn't even a guard on the premises. But the company is running tests in its recently built shale oil pilot works in Brea, California and it will make a decision this year to build or not to build a 7,000-barrel-per-day prototype commercial plant.

Union has its own land where the shale is rich and easily accessible through canyons. It has water rights methodically acquired over the years. It has, too, its own process featuring an ingenious device, known as a rock pump, which feeds the retort shale from the bottom. However, John Hopkins, who heads Union's new synthetic fuels division, is inclined to emphasize the ifs in the company's situation. "We don't want to do what Colony did," he says. "That is, say we are going ahead and then have to back down. *If* inflation abates and we can get some idea of what future costs will be, *if* we have some idea of what OPEC will do, *if* we can live with state and local regulations and taxes, we will go ahead, starting with a 7,000-barrel-per-day plant, then adding others up to 50,000 barrels per day. Our costs should come down after the first plant." His "ifs" are big.

Colony, up Parachute Creek from Union Oil, has emphasized again and again that its decision last October [1974] to postpone its plans—after the cost estimates more than doubled—to $800 million—does not mean it has abandoned them. "We expect to go forward, but on a different time schedule," says Hollis Dole, general manager of the project. Dole, however, does not say what the new schedule will be.

Superior Oil is apparently working on yet another

process—although it is keeping an even lower profile than Occidental.

So there is fierce competition in shale country right now to develop an economical, environmentally acceptable process. Considerable spoils in the form of licensing fees could go to the winner—if there is one. At the same time, work is proceeding on the four federal tracts that were leased by the Department of the Interior a year ago.

The lease itself stipulates that the companies must spend two years collecting base-line data. What this means essentially is that the oil companies have hired environment monitoring companies to count and catalog every living thing on the 5,000-acre tracts, to monitor continuously what is in the supposedly pristine air, and to test, also continuously, the quality of any water on the land. Ornithologists, terrestrial biologists, meteorologists all have their oars in.

Work on the federal tract is overseen by Peter Rutledge and his staff of fifteen. Rutledge is the area shale oil supervisor, appointed by the Secretary of the Interior and working in Grand Junction. The leases were meant to give the nation prototype shale development and, barring an Alaska-style tie-up in the courts, it seems clear that they will do that.

Beyond the prototypes, however, there are nothing but question marks. Will one of the processes prove cheap enough to make oil from shale competitive? Will the price of oil drop sharply in the world, pricing shale out of any conceivable market? Will the oil companies earn sufficient profits to enable them to experiment with far-out energy alternatives? Will inflation slow down enough to make the capital costs bearable? Or will that oil remain forever locked in rock?

"Big-scale shale oil will be developed," says a middle-aged executive of one of the projects, "but maybe not in my lifetime." Maybe not, but considering the economic and political importance of oil, it is a good thing that determined engineers are continuing their quest in shale country. Coal and nuclear energy may right now look like more practical alternatives to petroleum, but too much is at stake to ignore a potential treasure-trove like shale.

III. THE FOSSIL FUELS: OIL AND NATURAL GAS

EDITOR'S INTRODUCTION

Until the 1950s the United States was a net exporter of oil. But domestic use of petroleum products—both for fuel and as an ingredient in other products—grew so steadily that by 1973 the United States was buying 30 percent of its petroleum from abroad, mostly from Arab oil interests. As long as this Arab oil was cheap, there seemed to be no special need to reduce consumption or to find other sources. Furthermore, the United States had large supplies of various forms of energy—including petroleum—whereas other Western nations, Japan, and many developing countries, were forced to rely almost entirely on imported petroleum.

Thus, before 1973 few people noticed that anything like an energy problem existed in America. But when the Arab oil embargo abruptly cut off most importation of petroleum, the vulnerability of the United States was made obvious to all with startling force. When the boycott ended, oil prices began a dizzying dance upward. The United States spent $3.9 billion to buy foreign petroleum in 1972. Two years later the cost was $24 billion.

Because 63 percent of the world's known oil reserves lie beneath the sands of several Arab states, the Arab world has become the fulcrum on which energy discussions must turn.

The opening article in this section, from *Time,* puts the oil situation into historical and political perspective, focusing on the late King Faisal of Saudi Arabia, the monarch who mapped the strategy that catapulted the oil countries into international power and wealth. Faisal himself was shot and killed shortly after the article appeared, but his brother and successor has followed his policies closely. Saudi

Arabia remains the major force among oil-producing nations.

The next article illustrates a recurring difficulty in most discussions of energy: the lack of solid information. Julius Duscha writes of the inadequate data pertaining to petroleum and natural gas. Other articles in this compilation underline the truth of his observation: well-informed authors use sharply varying figures on reserves, forecasts, consumption, demand. Where energy is concerned, the statistics are very slippery.

How much oil is there? An article from *Fortune* discusses petroleum resources that the author, Sanford Rose, believes are overlooked in most predictions. He contends that better methods of secondary recovery can pump more oil and gas out of supposedly exhausted fields.

The next two articles, from *Business Week,* explore other aspects of the search for more oil and natural gas, two fuels that are frequently found in the same field. First is an article dealing with the problems and prospects of locating oil and natural gas off the continental shelf of the United States, and second is a piece that describes the worldwide search by oil companies for new fields.

An article from *Commentary* argues that the sharp increases in oil prices by the Organization of Petroleum Exporting Countries (OPEC) have unwittingly written an end to the era of oil. Country after country is rushing to develop alternative fuels. Thus, the author believes, the OPEC price rises have spelled the end of oil's dominance.

Of all the fuels now in heavy use, natural gas is the least objectionable in environmental terms. It does not pollute the air when burned, and it is transported more or less invisibly in pipelines. Unfortunately, this desirable fuel, which now provides about one third of total US energy needs, is not abundant. Proven domestic reserves have shrunk as use increased. In many states, natural gas companies cannot accept any new customers.

The last article in this section discusses the squeeze on

natural gas, and the arguments over state and federal rulings on pricing natural gas.

FAISAL AND OIL [1]

In every car and tractor, in every tank and plane—oil. Behind almost every lighted glass tower, giant industrial plant or little workshop, computer and moon rocket and television signal—oil. Behind fertilizers, drugs, chemicals, synthetic textiles and thousands of other products—the same substance that until recently was taken for granted as a seemingly inexhaustible and obedient treasure. Few noted the considerable historic irony that the world's most advanced civilizations depended for this treasure on countries generally considered weak, compliant and disunited. Now all that has changed, and the result has been a major economic and political dislocation throughout the world.

The change became dramatically apparent in 1974, a pivotal year that saw the decline of old powers, old alliances, old philosophies—and the rise of new ones. The West's belief in the inevitability of human progress and material growth was badly shaken as inflation spread oppressively across the world, several industrial societies tumbled into recession, and famine plagued a score of nations. There was a marked erosion in the wealth, might and cohesiveness of North America, Europe and Japan. In the developing world, forty or more countries with few natural resources fell increasingly into destitution and dependency. Meanwhile, a handful of resource-rich nations gravely compounded the problems and challenged the vital interests of the rest of the world by skillfully wielding a most potent weapon: the power of oil.

The Swiftest Transfer of Money in History

United in history's most efficient cartel, these nations exploited modern civilization's dependence on oil. Their

[1] From article in *Time*. 105:8-32. Ja. 6, '75. Reprinted by permission from *Time*, the Weekly Newsmagazine; Copyright Time Inc.

power came from the uniqueness of oil, an exhaustible and not quickly replaceable resource that has long been shamefully wasted by much of the world. Because oil is not usually found where it is most consumed, and demand for it is so great, it is the most widely traded commodity in world commerce as well as a highly volatile element in world politics.

Again and again, the cartel formed by the Organization of Petroleum Exporting Countries [OPEC] raised the price of oil until it reached unprecedented and numbing heights. The producing nations' "take" from a barrel of oil, less than $1 at the start of the decade, was lifted from $1.99 before the Arab-Israeli war . . . [of October 1973] to $3.44 at the end of 1973 to more than $10 at the end of 1974. The result is the greatest and swiftest transfer of wealth in all history: the thirteen OPEC countries earned $112 billion from the rest of the world . . . [in 1974]. Because they could not begin to spend it all, they ran up a payments surplus of $60 billion. This sudden shift of money shook the whole fragile structure of the international financial system, severely weakened the already troubled economies of the oil-importing nations and gave great new political strength to the exporters.

The beneficiaries of this transfer were a disparate group of oil-possessing Africans, Asians, Latin Americans and, most favored of all, Arabs, who provided two thirds of the petroleum exports and have more than three fifths of the proven petroleum reserves in the non-Communist world. One bleak, sparsely populated country is by far the world's greatest seller and reservoir of oil, and one dour, ascetic and shrewd man is its undisputed ruler. . . . [King Faisal was assassinated on March 25, 1975. His successor, King Khalid, continues the oil policies described in this article.—Ed.]

All the King's Spending and All the King's Plans

. . . [In 1974] Faisal's Saudi Arabia earned $28.9 billion by selling nearly one fifth of all the oil consumed by non-Communist countries. The king channeled part of these funds into a massive development program that aims at building factories, refineries, harbors, hospitals and schools for his 5.7 million people. . . . But all the king's spending and all the king's plans could not come close to using up Saudi Arabia's wealth. The new financial giant of the world, Saudia Arabia in 1974 stood to accumulate a surplus of about $23 billion—a potentially unsettling force in global finance.

Moreover, Saudi Arabia's new wealth is simply the most spectacular symbol of the rising fortunes of the OPEC nations. With their surplus of some $60 billion . . . [in 1974], they took in $164 million more each day and $6.8 million more each hour than, by best estimates, they can currently spend. At that rate of accumulation, the *Economist* of London calculates, OPEC could buy out all companies on the world's major stock exchanges in 15.6 years (at present quotations), all companies on the New York Stock Exchange in 9.2 years, all central banks' gold (at $170 an ounce) in 3.2 years, all US direct investments abroad in 1.8 years, all companies quoted on stock exchanges in Britain, France and West Germany in 1.7 years, all IBM stock in 143 days, all Exxon stock in 79 days, the Rockefeller family's wealth in 6 days and 14 percent of Germany's Daimler-Benz in 2 days (which in fact Kuwait did in November—though for that little country, the purchase represented all of 15 days of oil earnings). . . .

The New Reality of Arab Power

One of the causes of the West's woes is that for too long it underestimated the will and power of Faisal and other rulers of oil-producing nations to act together. The cries for higher prices had been rising for fifteen years, first from the Venezuelans and Iranians, then from the radical Arab leaders of Libya, Algeria and Iraq. Faisal, a conservative and a longtime friend of the United States, at first resisted—and then changed his mind because of US political and military support of Israel.

For many frustrating months in 1973, the king, and his spokesmen, warned the United States that unless it forced Israel to withdraw from occupied Arab territories and settle the Palestinians' grievances, he would slow down oil production. The State Department thought that the threat was hollow; President Nixon warned on television that the Arabs risked losing their oil markets if they tried to act too tough.

The Arab-Israeli war of October 1973 moved the Arabs to impose a reduction in oil output—and do much more. Within ten days after the Egyptians and Syrians had attacked Israeli-occupied territory, the Arabs and Iranians in OPEC—long derided in the West for their disunity—coalesced and raised prices from $1.99 to $3.44 per barrel. (The members of OPEC in order of . . . [their 1974] earnings are: Saudi Arabia, Iran, Venezuela, Nigeria, Libya, Kuwait, Iraq, United Arab Emirates, Algeria, Indonesia, Qatar, Ecuador and Gabon, which is an associate member. The United Arab Emirates is a federation of Abu Dhabi, Dubai, Sharjah, Ajman, Umm al Quwain, Ras al Khaimah and Fujairah.) A few days after that King Faisal led an even stronger move. Angered by the US military resupplying of Israel, the Saudis and the other Arabs embargoed all oil shipments to the United States and started cutting production. Very quickly their output dropped 28 percent. When the West made no response, OPEC realized its own strength and kept right on raising prices through 1974. . . .

So much foreign money washed into Arab oil-producing countries that ordinary statistics no longer made sense. Estimated gross national product per capita ran to $13,000 in Kuwait, $14,000 in Qatar and more than $23,000 in Abu Dhabi. But those figures did not reflect living standards because the quick cash has not had time to filter down to the people. Bureaucracies strained to figure out ways to spend at home. Kuwait expanded one of the world's most all-encompassing welfare states. To hold down food prices, most of the big oil producers subsidized imports of staples.

Office buildings, low-rent apartments and supermarkets rose almost everywhere. Some planners worried about keeping a work ethic going. Said a Saudi government minister: "We will have to be very careful not to spoil our citizens. Our people will have to deserve what they earn. We will furnish them with basic requirements, but nobody should live on charity."

The Europeans and the Japanese, umbilically dependent on the Middle East for respectively 70 percent and 80 percent of their oil, not only pressed their most modern technology on the Arab states but also granted them strong diplomatic support. Some European political leaders called for a new Euro-Middle East alliance, perhaps to replace the Atlantic Alliance. The French, responding to what they call the "new reality" of oil-based Arab power, were especially obsequious in their attentions. The Dutch, long outspoken defenders of Israel, fell silent in fear of Arab wrath.

The Aim: A Redistribution of Wealth

. . . Next to Faisal, the ruler who gained most from oil . . . [in 1974] was not an Arab but the "light of the Aryans," the shah of Iran. His country, the world's second largest oil exporter, quadrupled its petroleum earnings, to $20.9 billion. Impatient to industrialize and militarize, the shah pressed the construction of automobile and petrochemical factories, dams and hospitals, and ordered 70 F-4 Phantom jets and 800 British Chieftain tanks to bolster a mighty armed force. This swelling strength raised apprehensions among some Arab governments in the region and evoked new hostility—but also won new respect in Washington, where Iran is valued as an anti-Communist bulwark. Though much poverty and illiteracy hang on in Iran, the middle class is rapidly spreading and the gross national product is expanding at an astounding rate of 50 percent a year. The shah, who aims to turn Iran into "the Japan of West Asia," argued for price increases long before Faisal did, and he has been even more vocal than the Saudi king in urging that prices stay up.

Several other countries rose on petropower. Oil made Nigeria not only black Africa's wealthiest nation ($9.2 billion in earnings) but unquestionably its strongest political force. Indonesia, though still abysmally poor, is showing the first glimmerings of its potential as Southeast Asia's economic leader, thanks to oil exports. Oil-endowed Venezuela at midyear [1974] trebled its national budget, to almost $10 billion, to take account of rising revenues. The Venezuelans are expanding their state-owned steel industry in the Orinoco backlands, paying to educate thousands of future leaders at US universities and gaining great influence among Central American republics by promising them loans. Says Venezuela's President Carlos Andrés Pérez: "This is our opportunity to create a new international economic order."

A new order is the ultimate goal of the petrocrats. Their aim is to lead many of the third world nations in an economic revolution that is already bringing a radical redistribution of the world's wealth and political power. The transfer of riches to the oil producers has helped slow or stop the rise of living standards in many other countries—a development that has potentially grave social consequences. The steep economic growth that the industrial nations have enjoyed since World War II tended to soften social and economic inequalities because even the poor and deprived made visible progress year by year and could discern a brighter future. Now, if there is slow growth or no growth, demands for social justice will be more urgent—and harder to fulfill. Democratic governments will have to find ways to redistribute the existing wealth, or else face dissension and perhaps chaos.

The shah of Iran laid it on the line: "The era of terrific progress and even more terrific income and wealth based on cheap oil is finished." Henry Kissinger sees it another way. If high energy prices persist, he warns, "the great achievements of this generation in preserving our institutions and constructing an international order will be imperiled."

Inflaming Problems and Inflating Prices

The sudden, sharp rise in oil prices inflamed all sorts of problems, increasing government controls, intensifying nationalism and calling into question the future of free economies. People were gripped with the fear that events had overtaken their ability—or their government's ability—to cope. Otherwise sober men spoke of extreme solutions: repudiation of international debts, massive currency devaluations, the suspension of parliamentary government, even military intervention in the producing countries.

It was possible to blame too much of this malaise on oil. Many countries have long suffered from high inflation because they were living beyond their means for years. Particularly in the West's mass-consumer societies, the poor wanted to live like the middle class, and the middle class wanted to live like the rich. Demands piled up—for more goods, fatter wages, higher social welfare—and prices soared. Still, by best estimates, the rise in energy prices caused one quarter to one third of the world's inflation last year. As the price of oil increased, it kicked up the prices of countless oil-based products, including fertilizers, petrochemicals and synthetic textiles. To battle inflation, all Western nations clamped on restrictive budget and credit policies, causing their economies to slow down simultaneously for the first time since the 1930s.

The danger of a global recession grew because, as people spent more for oil, they had less money left over to spend on other things. The overall decline in demand reduced production and jobs. Because non-OPEC nations had to pay out so much for foreign oil, they moderated their buying of other imports; that slowed the growth of world trade, which has been a major source of international cooperation since World War II. The United States' relations with its allies also came under strain, and the West seemed without will or unity. For most of . . . [1974], Western European nations and Japan refused to follow the United States' call for a united front against the oil producers, essentially because European leaders considered the consumers' bargaining power too feeble.

The United States was a major oil exporter through the late 1950s, but then its own demands raced so far ahead of production that it now has to import more than one third of its supply. The nation's bill for foreign oil pyramided from $3.9 billion in 1972 to $24 billion . . . [in 1974]. The $20 billion jump meant that Americans either had to increase their foreign debts greatly or produce and export $20 billion more in goods and services—food, steel, planes, machinery, technology—to pay for oil imports. Unless the oil price comes down or the country sharply reduces its oil imports or substantially increases production, the United States will have to spend that extra $20 billion or more every year. This will drain off more of the nation's resources and build up trade debts that future generations will have to pay. In 1974 the rippling effects of rising oil prices contributed three or four percentage points to the US inflation rate of 12 percent. The oil rise, which Yale economist Richard Cooper called "King Faisal's tax," reduced Americans' purchasing power and consumption of goods as much as a 10 percent increase in personal income taxes would have done.

Nations that depend even more on OPEC fared much worse than the United States. Japan's $18 billion bill for oil imports was the biggest single factor in lifting its inflation rate to a punishing 24 percent, causing the first real postwar decline in economic growth. Inflation rates doubled in many Western European nations: to 16 percent in France and Belgium, 18 percent in Britain, 25 percent in Italy. To meet its trade deficit, Italy has borrowed more than $13 billion, incurring interest payments of nearly $1 billion a year. Prime Minister Harold Wilson says that the fivefold increase in oil prices aggravated Britain's worst economic crisis since the 1930s, and is severely testing the country's social and political fabric. Only West Germany, the Netherlands and Belgium ran trade surpluses.

For Europeans, life became a little darker, slower, chillier. Heating oil prices went up 60 percent to 100 percent, and thermostats were turned down. In the midst of a French conservation drive in October, President Valéry Giscard d'Estaing found his Elysée Palace dining room so cold that he lunched with Premier Jacques Chirac in the library by a crackling fire. Gasoline rose to $1.40 per gallon in West Germany, $1.72 in Italy, $2.50 in Greece. Electrical advertising signs were banned after 10 P.M. in France and during the daytime in Britain. In Athens, the floodlights illuminating the Acropolis were turned off. Throughout Western Europe, energy costs were a cause of the slump in sales of autos, houses and electrical appliances. Layoffs spread in those and other industries. Unemployment hit a postwar high in France. In Germany, foreign workers were being paid bonuses to quit and go back home to Spain, Turkey and Yugoslavia.

The Soviets benefited from what they accurately enough called this "crisis of capitalism." From their oil exports, mostly to the West but also to their East European allies, the Soviets earned $2 billion . . . [in 1974]. However, Russia will rapidly scrape the limits of its self-sufficiency if it is to meet plans to expand its petrochemical industry and treble auto ownership (to 9 million cars) by 1980. Soon the Soviets will have to restrict oil sales and greatly increase the preferential prices that they charge to their Comecon [Council for Mutual Economic Assistance] partners. . . . [In 1974] Poland reportedly had to buy a large amount of Libyan crude, at $16 to $20 per barrel. Strapped for hard currency to pay for oil from non-Communist sources, East Germany had to restrict the expansion of its plastics and textiles industries.

The poorest countries of Africa, Asia and Latin America were the worst hurt victims of the oil squeeze. Indeed, the developing countries' extra costs for oil . . . [in 1974] totaled

$10 billion, wiping out most of their foreign aid income of $11.4 billion from the industrialized world. In black Africa, only Nigeria has any big known reserves of oil, and Gabon, the Congo Republic and Angola possess some oil. For the other black African countries, the petrobill came to $1.3 billion . . . [in 1974]. Development plans were stymied because so much money was drained off for oil. Drought-induced hunger became worse, in part because those countries could no longer afford as much gasoline to run their tractors, or fertilizers to nourish their fields. Inflation raced at rates averaging 45 percent.

India suffered more than any other nation. Its oil import costs hit $1.6 billion, up fivefold in two years, leaving it little money to import food and fertilizer, machines and medicine for its hungering millions. Pakistan's plight was almost as critical; its imports of oil and fertilizer topped $355 million. Sri Lanka's rice farmers had to pay 375 percent more for fertilizer; they reduced their buying so much that the rice harvest fell almost 40 percent below expectations.

The poorest countries—those with scant resources to finance their needed imports—descended into a new category, now known as the fourth world. The old third world became a more exclusive, OPEC-led grouping, limited to those nations that are exploiting their rich mineral or agricultural resources. Emboldened by the oil producers' success, many other third world countries tried to create their own price-fixing cartels for copper, iron ore, tin, phosphates, rubber, coffee, cocoa, pepper and bananas. Their leaders talked of "one, two, many OPECs." The grand plans generally failed because members have lacked the cohesiveness to make them work—so far. But the new importance of raw materials moved some big producers to raise prices unilaterally. Jamaica, for example, abrogated contracts with companies and lifted the government take for the country's bauxite by 700 percent.

In sum, the world has entered an era in which natural resources will count for much more than before, conservation will gain a premium over consumption, and more attention will be paid to exploiting resources than curbing pollution. All this will bring many changes in life styles: slower gains in real purchasing power, stricter controls on energy use, smaller cars. It remains to be seen to what extent the changes will be accepted by such disparate forces as labor unions, auto manufacturers, and consumer and environmental groups.

The Case For—and Against—Increases

With passion, the oil producers defend their price increases on the ground that it is high time that the producers of raw materials get a fair shake from the richer industrial nations. Essentially, these are the oil producers' arguments:

In the past, the industrial countries grossly exploited the oil-producing countries. For too long, the terms of trade were stacked against the materials producers. While they were forced to pay ever inflating prices for their machines, medicines, food and other goods bought from the West, the developed countries not only imported oil at low, stable prices but also built industrial and consumer booms on it. Now the oil producers must build their own industries, both to get a more equitable share of the world's income and to insure themselves against the day when their petroleum resources run out. Furthermore, by keeping prices high, the producers are really doing the rest of the world a favor by forcing both energy conservation and the search for alternative resources.

The rise in oil prices, the producers go on, should not get all or even most of the blame for inflation, slow growth and balance of payments problems, which have deeper roots. Says Kuwait Oil Minister Abdel Rahman Atiqi: "Why should we be responsible for helping the United States, for instance, solve its economic problems? When our Arab lands were impoverished and our oil was being sold at giveaway prices, what assistance did the United States give us?" . . .

The OPEC nations cannot accurately argue—either in

terms of economics or "fairness"—that the sharp rise to $10.12 per barrel is needed to make up for the recent inflation in the price of goods that they buy in world trade. John Lichtblau, a leading US oil consultant, notes: "Since 1960, the UN index of world export prices of manufactured goods has risen 86 percent and the Saudi government's revenue on each barrel of oil has risen 1,136 percent. Since 1970, world export prices have risen 55 percent and the OPEC governments' income on each barrel has gone up 955 percent."

The high prices will certainly discourage oil waste, but the producers have an exaggerated fear that they will soon run out of what the shah calls "this noble product." The Middle East's proven reserves have risen every year since records were kept and have doubled since 1959, to some 350 billion barrels. Saudi Arabia alone has proven reserves of 132 billion barrels—enough to keep producing at current rates until the year 2018—and some experts reckon that the real total could be four times as great.

Nobody knows what would be a "fair" oil price, but logically it should bear some relation to the cost of primary production. That cost ranges downward from $2.50 or so per barrel in the United States to 60 cents in Venezuela and 12 cents in Saudi Arabia. The price should also have some market relationship to the price of alternative energy sources, which many authorities think would be economically feasible when oil sells at $7 or more per barrel. But with the latest round of oil price increases last month, the OPEC governments will collect $10.12 on a barrel. By contrast, the international companies earn 20 cents to 50 cents per barrel in return for all the work, risk and investment that they undertook to find and pump that oil.

Thus, instead of the elusive terms of fairness, the argument is perhaps best couched in terms of ultimate self-interest. The oil producers may well be setting a dangerous precedent, for themselves as well as oil users. By exercising monopoly muscle as a group of nations, the cartel may be creating a world in which prices are neither fair nor free but fixed by raw economic power. Considering the fact that oil is about all they have to bargain with, that kind of world could eventually be dangerous for OPEC's members. The oil producers quite frankly say that they expect the living standards of Western industrial countries to grow at a slower rate for the immediate future, and they cannot be expected to weep over that. But by forcing the change so suddenly, without giving the oil importers a chance to adjust gradually, OPEC runs the risk of wrecking the world economy—and that, OPEC spokesmen themselves have admitted, could only hurt them.

The Companies' Rich Past and Questionable Future

In all this, the role of the oil companies is growing weaker. The companies not only discovered and developed the oil but also put up billions of dollars to build rigs, pipelines, refineries and harbors. They have done so for more than forty years, since long before the Saudis had much interest in oil, let alone the means to exploit it. The first prospectors—from Standard Oil of California—went to Saudi Arabia in 1933 and brought in the first well in 1938. They and later prospectors had a rugged frontier existence, living in tents and huts, relying on an 11,000-mile-long logistics line from the United States, and coping with desert sand, burning heat and loneliness. In the late 1930s and early 1940s, they were joined by Exxon, Texaco and Mobil to form the Arabian American Oil Company [Aramco]. Oil prices were relatively low—$1.40 to $2 and the governments' take ranged from 20 cents to less than $1 a barrel—because Middle East production costs were modest, oil was in surplus in the world, and the producers' governments were weak and disunited. Company earnings were huge. When supplies tightened and producers began to get together in the late 1960s, the governments' split of production profits rose from 50–50 to 67–33. Even before the price rises since 1973,

Middle East governments profited nicely from oil; Saudi Arabia's take from 1965 to 1972 totaled $10 billion.

The OPEC countries have shrewdly turned the companies into scapegoats, blaming their high profits for the high retail prices. Indeed, in this year's first nine months, profits of the five biggest US international oil companies jumped anywhere from 38 percent to 70 percent. But much of this gain was due to an unusual circumstance: OPEC's price rises triggered an automatic increase in the value of the huge stocks of oil that the companies held in tank farms and on tankers. The companies will not get those one-shot "inventory profits" in the future, unless OPEC again raises the price. As for relative earnings, the five companies' profits rose from $5.3 billion in the twelve months before the embargo and big price rises, to a steep $8.2 billion in the twelve months following; but the OPEC governments' revenues swelled from $22.7 billion in 1973 to $112 billion . . . [in 1974]. The companies' earnings will probably decline . . . [in 1975] because their costs are going up while oil demand is going down.

The Danger of Rising Surpluses

The companies, in fact, were among the biggest losers of 1974. The four US partners in Aramco had to agree late in the year to sell their remaining 40 percent ownership to Faisal's government. It will pay the partners $2 billion for almost all their facilities, a price that the Saudis can meet with less than one month's oil earnings. The Saudi takeover will move Kuwait, Qatar, Oman and the United Arab Emirates to nationalize the last of the Western oil operations in those areas, probably . . . [in 1975]. The companies will become mere agents, selling technical and marketing services to the governments for a fee.

The major companies' future is uncertain as they will face competition for markets from the oil countries' state-owned companies. Some national producers want to squeeze the private oil companies because they are viewed as competitors. Mani Said Utaiba, petroleum minister of the United Arab Emirates, complained: "These profits are being used by [the companies] to find alternative sources for our oil. They are investing on a huge scale in the Arctic and the North Sea. This we will not accept."

The oil crisis promises to shake the world for at least another five years or longer. It will take that long for importing countries to develop alternative energy sources and more petroleum in nations outside OPEC. Oil will be flowing in from Alaska by 1978, but the total—600,000 barrels a day at first, 2 million barrels a day by 1981—will not free the United States from the need for foreign supplies. Britain and Norway are each expected to be pumping 2 million barrels a day from deep below the North Sea by the early 1980s. But the rest of Europe, as well as Japan and the fourth world, will still depend on Middle East oil, above all from the country that has most of it: Saudi Arabia.

Moreover, if . . . [the oil countries] hold prices up, the rest of the world could encounter such compounded problems that 1974 would be remembered as an easy year. With oil at $10 a barrel, OPEC would charge the world another $600 billion in the next five years. To pay the bill, the 137 nations outside the cartel would have to deliver one quarter of their total exports to OPEC's elite thirteen countries. It would be impossible for the oil importers to transfer so much of their production—or for OPEC nations to absorb it all. The most frightening figure for the future is that OPEC nations stand to accumulate payments surpluses of $250 billion to $325 billion by 1980, and the rest of the world would run up exactly that much of a deficit. For the countries that have them, surpluses create huge purchasing—and political—power. Conversely, deficits usually lead to recessions, devaluations and decline.

Both the surpluses and the deficits will drop when the OPEC countries expand their buying, lending and investing abroad. In stepping up their domestic development plans, they will have to enlarge their imports. This can be accomplished fairly easily by seven of the OPEC members: Iran, Venezuela, Indonesia, Iraq, Nigeria, Algeria and Ecuador. They have relatively big populations and much poverty—hence much need for internal development. The huge problem is that six other, lightly populated Arab states—Saudi Arabia, Libya, Kuwait, Abu Dhabi, Dubai and Qatar—are collecting far more money than they can possibly spend. These six, embracing only 9.3 million people, earned $54.7 billion from oil . . . [in 1974]. For all their industrialization and social welfare, their military and foreign aid, they can dispose of only a fraction of that total, leaving a combined surplus of $38 billion.

Naturally, the Saudis are piling up the biggest surpluses. At present prices and production levels, they will collect a staggering $150 billion over the next five years. But they will be unable to buy or build fast enough to use up even one third of their oil money on domestic development. By 1980, they stand to have well over $100 billion in surplus—to lend, give away or invest in foreign countries. . . .

Conserving to Crack the Cartel

. . . Even with the best of recycling, the importing nations will be vulnerable. Says Walter Levy, the world's leading oil consultant: "The world economy cannot survive in a healthy or remotely healthy condition if cartel pricing and actual or threatened supply restraints of oil continue." In many ways, Western democracies face a wartime-like crisis, but until lately they have reacted as they did during the 1939–40 "phony war." Only by cooperating among themselves can the importers counter the cartel's control over their destinies. Recently they have begun to make tentative moves to accomplish three necessary things: conserve energy, develop new sources and stockpile oil in case of another embargo or cutback.

In November [1974], ministers from the United States, Canada, Japan, all members of the Common Market (except France), four other European nations and Turkey signed an agreement to form the International Energy Agency [IEA], which Henry Kissinger had proposed. Provided their legislatures approve, each member would build up a stockpile of oil equal to ninety days of imports; if any OPEC members embargo oil or reduce shipments, the IEA nations would reduce consumption and later share what they have with one another. . . .

The Western nations will have no real bargaining strength until they show that they are taking strong measures to conserve. By significantly reducing demand, the big buyers of oil might force OPEC into production cuts that some cartel members may eventually find intolerable. Cutbacks would be particularly rough for Iran and Iraq, both of which plan substantial production increases in the next few years to finance their grand development programs. Rather than reduce output, other populous countries with ambitious development schemes—Nigeria, Venezuela, Indonesia—might be tempted to buck the cartel by selling below the fixed price. Ecuador, which badly needs development money, is already in some trouble. High prices have cut demand for its oil by one third since 1973.

At very best, however, the State Department reckons that OPEC would not break up for another two to four years—and probably not even then. It has not been at all damaged by a world oil surplus of 1 to 2 million barrels a day, which has shown up because high prices reduced consumption last year. In the non-Communist world, consumption fell from 48 million barrels a day in 1973 to 46.5 million barrels . . . [in 1974]; in the United States, it declined from 17 million barrels to 16.2 million barrels. Partly in response, OPEC is now producing at 20 percent below capacity with no visible problems. Again, it is Saudi Arabia that holds the key. The country has accumulated so much money that it could stop production for two or three years and still have more than enough cash to import food, provide free medical care and education, finance new industry and subsidize other Arab nations. But unless and until the industrial nations get together, much of the non-Communist world could not long function without Saudi Arabia's 8.5 million barrels per day. As Saudi Arabia's Harvard-educated Oil Minister Ahmed Zaki Yamani . . . [said]: "How much can the consumers reduce consumption? By 10 percent? And how much can the producers reduce without financial pain? By at least 33 percent—minimally. The people who ask for a price reduction of $2 to $4 are simply not being realistic."

Even so, the consumers must conserve to show OPEC that they are serious and to hold down their payments to the cartel. Kissinger has urged that they hold their oil imports essentially flat over the next decade. For the United States, that would mean a decline in the annual rate of increase in energy from 4.3 percent in the past ten years to 2 percent or 3 percent in the next decade. The Trilateral Commission has called for limiting the annual growth in energy use during that period to 2 percent in the United States and Canada, 3 percent in Western Europe and 4 percent in Japan. Certainly the United States can and must lead the way by making the severest cuts because it wastes so much energy. A nation that has one twentieth of the world's population should not expect to go on burning one third of the world's oil.

Through taxes and other mandatory measures, the United States could switch from profligacy to a new conservation ethic. The remedies are well known. Much energy could be saved by increasing federal taxes on gasoline, clamping a steeply graduated tax on heavy, thirsty cars, pumping many more millions into mass transit, and granting tax credits for purchases of building insulation. In addition, the United States could and should expand its domestic supplies of energy by increasing the capacity of the Alaska pipeline, opening the Navy's petroleum reserves in California and Alaska, encouraging offshore drilling, liberalizing controls on the strip mining of coal (but adding guarantees that the lands would be reclaimed) and allowing natural gas prices to double or more.

The Perils of Military Intervention

Beyond conserving energy and recycling OPEC's money, the oil importers have no feasible weapons against the cartel. A trade war against OPEC would fail. If the United States, for example, embargoed its shipments of food or machines to the oil producers, the Soviet Union and other countries would be eager and able to fill the gap.

Military intervention could be extremely risky. There is always the danger that the Soviets would step in on the side of the Arabs—or extract a high political price from the West for staying out. Pipelines might be vulnerable to sabotage, though captured oil fields could be fairly easily protected. In any event, US authorities condemn the wave of fantasizing about oil wars as "highly irresponsible." Military intervention, says a Washington policy maker, would be considered "only as absolutely a last resort to prevent the collapse of the industrialized world and not just to get the oil price down."

The United States would be forced to use its "military option," however, in the case of any clear and immediate danger that Saudi oil would fall into hostile hands. There is concern in Washington that in several years an extremist force might try to grab control. . . .

A settlement with Israel would not itself lead to a price reduction. The non-Arab nations—Iran, Venezuela, Nigeria, Indonesia—though not part of the conflict, still want to maintain or increase prices. Yet marked progress toward peace on terms acceptable to the Arabs is absolutely essential before prices can soften; the Arabs will insist on that.

On the other hand, if war erupts anew, the Arabs might embargo either the United States or all Western nations. Says Saudi Interior Minister Prince Fahd, 53 . . . [Faisal's brother and the new crown prince]: "We would hate to impose another embargo. But in a war, when you feel you are in danger of dying, you may do anything. If war breaks out again, it will be not only the Arabs and Israelis who are damaged, but the world as a whole." If Western Europe were embargoed now, it would draw down its stockpiles (good for sixty days or more in each country), buy oil from non-Arab countries and probably go to immediate rationing. It might well hold out for six months without serious discomfort. Quite probably, however, Europe and Japan would put extreme pressure on the United States to halt military aid to Israel. Or, if threatened by complete economic breakdown and perhaps social upheaval, some Western nation or nations might intervene in Middle East oil lands. In any case, there is virtual consensus among Western policy makers that Israel must give up almost all of its 1967 conquests and accept a homeland for the Palestinians. Otherwise, wars are likely to continue, and Israel cannot win the last round against 120 million Arabs enriched and armed by oil money.

The Only Alternative: Interdependence

One ray of hope in the oil crisis is that the two sides at least will begin to talk with each other in 1975. The Middle East producers have long called for a summit meeting with the oil importers from the West and the developing world. The French have strongly favored a conference. Kissinger has held out for a delay until the consumers are more firmly united, fearing that countries that are deeply in debt and heavily dependent on oil imports would easily bend to OPEC's bidding. At Martinique . . . [in 1975], President Ford and French President Valéry Giscard d'Estaing struck a compromise calling for a series of meetings: first a general feeling-out between OPEC and the consumers, then a number of meetings among consumers to work out their common position, and finally a tripartite summit, probably . . . [in] autumn [1975].

At that summit, OPEC leaders want to discuss not only oil but also the prices of other products. They aim to get an "indexing" agreement under which their oil prices would go up from the already high base as the prices of their own imports rise. Says Kissinger: "The best thing that can happen next year—and in fact I think the best will happen —would be that we would achieve consumer solidarity and then have a conference with the producers. That, together with energetic conservation measures and energetic development of alternative resources, may lead perhaps to a lowering of the oil price in return for long-term stability of the price. And at a lower price level, we would be prepared to consider indexing."

A most positive step would be for oil producers and consumers to seek common and reciprocal interests going far beyond energy. The producers should be given greater responsibilities and more high offices in international councils. For example, they should get far more than the 5 percent of the voting strength that they now have in the World Bank and the International Monetary Fund. This would give them a larger voice in setting international monetary policies, which they deserve, and would also oblige them to put up quite a bit more than the 5 percent that they now give to underwrite those groups. The producers have been increasing their foreign aid fairly rapidly, but they probably should give much more in grants, low-interest loans and concessionary prices to the neediest countries. . . . [In 1974] OPEC members made aid commitments totaling $9.6 billion and actually disbursed $2.6 billion in gifts, concessionary loans and other aid—roughly half of it to Egypt, Syria and Jordan.

Whatever devices are created to put OPEC capital to work in the rest of the world, the Western countries should help the oil producers build up their own agriculture and industries. Faisal notes, for example, that his rich country badly needs industrialization. To help prepare the producers for the day, however distant, when their oil runs out, the West should also join them in developing alternative forms of energy and should send technology and experts to OPEC countries. Fast development is inevitable in the oil countries, and it will help work off their surpluses by spurring their imports. For their part, OPEC members may lend or invest some of the huge sums of capital that oil importers will need to develop energy supplies from the atom, from shale and sands and, probably many years from now, from the sun and wind.

In the difficult decade ahead, the best hope is that all sides will realize that they are really interdependent—for resources, technologies, goods, capital, ideas. The old world of Western dominance is dead, but if the oil powers try to dominate the new world of interdependencies, the result will be bankruptcies and deflation in the West, and even worse poverty and hunger in the have-not developing countries.

The oil producers, who talk a great deal about past exploitation and their future aspirations, might consider the implications for themselves of the havoc that their monopoly pricing is causing the rest of humankind. The oil consumers, who are the victims of that upheaval, would do well to ponder with more sympathy the OPEC countries' deeply felt desire for a larger share of the world's wealth. In this great global clash of interests, it is time for both sides to soften their anger and seek new ways to get along with each other. If sanity is to prevail, the guiding policy must be not confrontation but cooperation and conservation.

OIL: THE DATA SHORTAGE [2]

Last October 12 [1973], when the Arab-Israeli war was beginning to raise fears of an Arab boycott of oil sales to the United States, White House petroleum expert Charles Di Bona said American imports of Arab oil had been averaging 1.2 million barrels a day. Eight days later, on October 20, the Nixon Administration was talking about US depen-

[2] From article by Julius Duscha, director of the Washington Journalism Center. *Progressive:* 38:23-5. F. '74. Reprinted by permission from *The Progressive,* 408 West Gorham Street, Madison, Wisconsin 53703. Copyright © 1974, The Progressive, Inc.

dence on Arab oil totaling 1.6 million barrels daily. Four more days passed, and on October 24 the Administration estimate escalated to 2 million barrels. By October 30 the figure was up to 2.5 million barrels, and then, early in November, the Defense Department, never to be outdone even by the White House, proclaimed a 3-million-barrel-a-day US dependence on Arab oil.

When estimates varied within a month all the way from 1.2 million to 3 million barrels, it was obvious that no one in the federal government knew with any accuracy how much Arab oil was being used in the United States. The problem isn't that federal oil economists and statisticians can't count; it is rather that there are no reliable figures for them to count.

The statisticians do know that last September shipments of oil directly from the Middle East to the United States averaged 1.2 million barrels a day. However, US petroleum experts believe—but have no way of knowing for certain—that as much or more Arab oil also got here after being transshipped or processed somewhere along the way, usually in the Caribbean.

Once crude oil from the Middle East or from any other area enters world commerce, its route is difficult to follow; oil bears no serial numbers or other distinguishing marks of manufactured goods. The federal government could be forgiven the wide variances in its estimates of Arab oil imports were it not for the fact that the Arab import statistics are symptomatic of the problem with most petroleum figures.

As everyone concerned with the oil crisis, from energy czar William Simon [now Secretary of the Treasury] to Senator Henry M. Jackson, has quickly come to realize, the United States has all too few solid figures on oil and natural gas, and for what statistics are available the government is almost wholly dependent on the oil industry itself.

Just as Arab import figures have jumped all over the place, so have Administration estimates of the dimensions of the overall US oil shortage this winter. Early in November

[1973], when President Nixon first talked about a serious shortage, he put the likely daily shortfall at 3.5 million barrels. Since then estimates by experts within the Administration and on Capitol Hill have ranged all the way down to 1.6 million barrels; Simon came out at 2.7 million barrels at the news conference late in December at which he outlined the Administration's standby gasoline rationing plans.

Whether the oil statistics are proclaimed by Simon's Federal Energy Office or by Jackson's Senate Interior Committee, almost all of the figures originate with the American Petroleum Institute [API], the trade association of the major US oil producers.

The API's basic statistical document is its *Weekly Statistical Bulletin,* which purports to keep track of the output of US refineries, available supplies of refined petroleum products, exports and imports, production of domestic oil wells, and gasoline consumption. Of these figures, the only up-to-the-minute ones are for refinery production and available supplies, and even these must be viewed with some skepticism.

They are, first of all, industry figures supplied on a voluntary basis, with no independent check on their accuracy.

Every Wednesday, refinery production and available supply figures are made public for the previous week, and even oil industry critics agree that the statistics could hardly be produced much faster and still be put together with reasonable accuracy and cost.

Import and export statistics are based on United States Customs Bureau figures, with which no one quarrels much, except that the figures give no hint of the country where the crude oil was produced, or any indication of US oil being exported and then reimported under more favorable prices.

As for the domestic crude oil production figures, they are estimates based on the most recent statistics from the Texas Railroad Commission and the other regulatory agencies that monitor production in the thirty-one oil-producing states. The official state figures lag months behind, as do accurate gasoline consumption figures based on state and federal tax figures.

Once fuel oil, gasoline, and other petroleum products have left refineries, bulk terminals, and pipelines, neither the API nor any government agency keeps track of the products. How much gasoline is generally in the supply pipelines beyond the primary storage facilities of the refiners, the bulk terminals, and the big pipelines? How much No. 2 heating oil is in supply lines? How much oil and gasoline is kept in storage, or perhaps being hoarded, by large corporations and other heavy users of petroleum products? A recent New York *Times* survey indicated that major users of oil products have been building up their supplies, but there is no reliable statistical information available.

Oil-pricing information is even scarcer. Not until last fall [1973] did the United States Bureau of Labor Statistics [BLS] start collecting enough information on retail gasoline prices to come up with reasonably accurate figures on gasoline price trends throughout the nation. And there simply are no reliable figures on crude oil production costs, refinery costs, and costs on down the petroleum supply pipelines. There is more information on natural gas because most of its production and prices at the wellhead are still regulated by the Federal Power Commission [FPC], but the Nixon Administration has called for ending such regulation. And even data published by the FPC on natural gas production are heavily dependent on unaudited reports from the industry.

What government oil statistics there are come largely from the Interior Department's Bureau of Mines and Office of Oil and Gas. These are unchecked industry figures issued only monthly. To keep up with the fast-moving oil and gasoline situation, energy director Simon and his aides rely, however, not on the ponderous Bureau of Mines and Office of Oil and Gas, but rather on the API weekly bulletins. For it is true, as John E. Hodges, director of API's Division of Statistics and Economics, says, that "if the government didn't have our weekly statistical bulletin, it wouldn't have anything."

Oil is by no means the only industry for which audited statistics are hard to come by. American corporations jealously guard their figures—in part because of fears that competitors will use them to exploit market situations and in part because of the traditional feeling that the government (and the public) have no right to pry into the affairs of private corporations, no matter how big or dominant they may be. Most Census Bureau figures, as well as statistics compiled by other government agencies, are based on unchecked material submitted by industry sources. BLS and population figures are, of course, the government's own.

But if oil production and supply statistics cause problems, figures on oil and natural gas reserves are even more difficult to weigh and interpret. The API and the American Gas Association [AGA] are the only sources for systematic compilations of what oil men call proved reserves. Such reserves are defined as oil and gas in the ground that can be efficiently recovered under current economic conditions in the industry, which means that reserves increase as oil prices go up.

Every spring API, AGA, and the Canadian Petroleum Association jointly publish a detailed compilation of crude oil and natural gas reserves in the United States and Canada. The figures are based on surveys made by geologists and others employed by oil and natural gas companies. It is generally assumed that the industry has a vested interest in underestimating reserves, but government oil experts cannot even agree on that. A recent Federal Trade Commission study concluded that natural gas reserves were overestimated, while an FPC study showed that gas reserves were being underestimated in at least one rich field.

"I wouldn't argue one way or another about the reliability of the oil and natural gas reserve figures," a Senate staff aide who has been immersed in the quarrels over oil statistics observes. "There's just too much guesswork in estimating reserves. The only thing that's clear to me is that we still have an awful lot of oil in the world, and are likely to find a lot more."

However unreliable figures on US oil and natural gas reserves may be, there is general agreement even among oil industry economists that statistics on foreign reserves are hopeless. API's Hodges says flatly that there are no good figures on overseas reserves.

The United States thus finds itself groping for facts as it faces what appears to be the most serious fuel shortage in its history. But are things as bad as the Nixon Administration has made them seem? There is wide skepticism in Washington and elsewhere in the country—not only because of Nixon's personal credibility crisis, but also because of the uneasy feeling so many people have that the oil companies and not the Nixon Administration are in charge of whatever crisis there may be.

The oil industry certainly is in charge of whatever statistical evidence there is to back up the crisis atmosphere. And the industry obviously does not want the government to get into the information-gathering business. Industry lobbyists fought against inclusion of a government fact-gathering clause in the energy legislation that was shoved aside in the pre-Christmas congressional adjournment rush. But Simon's energy office has set up a committee to review oil and gasoline statistics, and Simon himself told reporters at his late December [1973] news conference: "You're darn right we're going to get better figures."

Senator Gaylord Nelson has presented the Senate with a plan to establish a Bureau of National Energy Information in the Commerce Department to collect and verify statistics involving all aspects of energy. "We have failed to manage energy because we have failed to manage energy information," Nelson says. "We are sitting in the dark because we have been making our energy policy in the dark."

Beyond the energy crisis [Nelson continues] the basic premises of this legislation are, first, that the power of giant corporations over the quality of life has become so great that such corporations must now be regarded as if they were governments, for govern they do; second, that governments—including corporate governments—derive their just powers from the consent of the governed; third, that consent, to be meaningful, even to be real, must be informed consent; fourth, that the free exchange and availability of industrial as well as political information are therefore the lifeblood of a free society; and fifth, that the Congress has no higher duty than to provide channels and mechanisms for the exchange and availability of information about the holders and uses of governing power.

It seems all but certain that out of the energy crisis will emerge audited government figures on the facts of oil and natural gas. It would be useful to know as much about the oil industry as advertisers know, for example, about television audiences. The fact that more information is available about how to influence television viewers than about oil says much about the haphazard ways of the American economy and the relationship of the government to it.

OUR VAST, HIDDEN OIL RESOURCES [3]

That continuing dialogue about energy shortages, and their effect on the US economy, has suffered from what might be called the "middle period void." Some people tend to focus on the undeniable miseries of the short-term situation. Others focus on the dramatic long-term possibilities for moving to an era of abundant energy—an era, beginning around 1985, in which the United States might be a substantial exporter of fuels.

But what about that murky middle period—say, 1975–85? Will economic growth in these years be hobbled by shortages of energy? Will shortages continue to force prices up or, at least, keep them at unacceptably high levels? Must the United States be dependent on an uncertain supply of im-

[3] From article by Sanford Rose, an editor at *Fortune*. *Fortune*. 59:104-7+. Ap. '74. Reprinted from the April 1974 issue of *Fortune* magazine by special permission; © 1974 Time Inc.

ported energy in these years? For most Americans, it seems clear, the middle period is not yet in focus.

Among those who are inclined to optimism about this period, most are putting their faith in coal. And, undoubtedly, increased coal output can help us to protect our economy from the vagaries of Middle East politics. But for a variety of reasons—principally having to do with our environmental concerns—it would be highly desirable to minimize our dependence on coal.

How to Set Off an Explosion

And the fact is that we can get a good deal of the energy we need in this middle period from the domestic oil and gas industry. In fact, the industry is capable of a phenomenal explosion of output within the next three years. To set off this explosion, all we have to do is shed some large misconceptions about the economics of oil supply and about the geology of oil reserves. The "we" in this case includes a fair number of oil producers and government regulators.

One widespread misconception has to do with the size of domestic oil reserves. The oil industry says that it has "proved" reserves of about 37 billion barrels. Since annual output totals over 3 billion barrels, the reserves would seem to represent only about eleven years' output at present rates of production. But proved reserves seriously understate the amount of oil that is potentially available. According to the definition given by the American Petroleum Institute, proved reserves "are the estimated quantities of crude oil recoverable under existing economic and operating conditions." Since conditions have manifestly changed in the past year, so should the figure for reserves.

But the official figure has not changed, and it is clear that our potential for oil production is now considerably higher than the industry admits. If one turns from the API's "proved reserve" concept to a simple "how-much-is-out-there?" concept, the totals change dramatically. Over the years, US oil producers have discovered about 430 billion barrels of oil in the United States. Approximately 100 billion barrels have been extracted thus far, leaving 330 billion still in the ground.

Some of this oil is in abandoned fields; some of it—including all the proved reserves—is in currently active fields. The new higher prices for oil make it profitable to attempt recovery of perhaps half that remaining 330 billion barrels. In other words, potentially recoverable reserves now represent over *fifty* years' supply at present rates of production.

This calculation, furthermore, includes only oil that has already been discovered—and we are bound to discover more. The United States Geological Service estimates that the lower forty-eight states contain somewhere between 575 billion and 2.4 trillion barrels of oil, some of it onshore and a good deal more on the continental shelf—for example, in the Gulf of Mexico and off the Atlantic Coast.

A Question About Capacity

Even if the oil is there, will we be able to get it out quickly enough? After all most oil wells in the United States are now operating at maximum capacity; many of the oil companies claim that they have their hands full just holding production at current levels. But capacity in the oil business has two quite distinct meanings. One refers to the amount of oil that can be produced in a given oil field with available supplies of equipment and existing pipelines. The other refers to the amount of oil the regulatory authorities will let the companies produce.

Many kinds of oil field equipment are unquestionably in short supply. Producers, particularly the small operators, report difficulty in obtaining drill pipe, oil well casing, and mud pump valves. Some producers complain that the steel companies are filling only about 10 percent of their equipment orders. But the steel mill capacity for meeting orders can be made available.

The steel companies just have not been making enough

of the "tubular goods" for the domestic market. . . . Price controls encourage the companies to export these products. Controls on steel products are likely to come off . . . [they did in 1974]. There is virtually no one in the oil business who suggests that current equipment shortages cannot be remedied within the next few years, provided the steel companies are given the price incentives.

Martial Law in Texas

A much more formidable obstacle to expanding oil output resides in the attitudes of the state authorities. The fact is often lost sight of in discussions of oil industry economics, but most oil production in the United States is under the control of state regulators. And the regulators have historically practiced at least three kinds of control on output.

One kind dates from those happier days when oil was in abundant supply. When the supply of oil exceeded the demand at "reasonable" prices, many states restricted output per well in order to prevent market disruption. The prevention of market disruption has long been a kind of unofficial religion for the regulatory authorities of several oil-producing states.

In a sense, they got religion forty-four years ago. In 1930 the price of oil hovered just above $1 a barrel. Then the giant East Texas oil field—still the country's largest—was discovered, and the price of oil tumbled to 10 cents a barrel within months. The Texas authorities declared martial law and imposed output restrictions at gunpoint.

Until quite recently, the restrictions were often defended largely on the grounds that they were protecting small business. It was held that unless output was curtailed, the large, vertically integrated companies would extract as much as was needed to supply their own refineries. The smaller companies, which did not have their own refineries, would then be unable to find outlets for their crude oil and would be forced out of business. Thus a policy plainly designed to maintain prices at high levels was insistently represented as a benefit for the "little man."

Beginning in 1972, when supplies of crude oil tightened, the argument became moot. Restriction of output for avowedly economic reasons ended. But restrictions for "conservation" purposes are still in effect. Texas, which accounts for more than a third of all US production, retains the most elaborate machinery for output restriction.

There are nearly 8,000 underground petroleum reservoirs in Texas, containing about 177,000 oil wells. Production from any given reservoir may be confined, by law, to its "maximum efficient rate [MER]." A field's MER is defined as the highest rate of production that can be sustained over a long period of time without reservoir damage or significant loss of ultimate oil and gas recovery.

As a reservoir becomes depleted, the water or gas pressure forcing the oil toward the wells generally begins to decline. Eventually, the pressure may fall to a point at which further production becomes exceedingly difficult and the reservoir may be abandoned—even though much, perhaps most, of the oil is still not recovered. It is this prospect that underlies the MER concept.

The concept has become a source of endless warfare between the companies and the state authorities. The companies usually think that they should be allowed to produce more; the state usually thinks it would be dangerous to raise the limit. In the last few years, controversy has centered on nine huge oil fields in Texas, including the giant East Texas field. The fields are today operating at MERs that many companies believe are far too low.

Only a Matter of Days

The companies contend that if they were allowed to increase output in these fields, they could get about 600,000 more barrels of oil a day. That total alone is equal to nearly two thirds of the average daily shipment of Arab oil to the United States during 1973. Moreover, extra production from most of the nine fields could come on-stream within a matter of days.

Not every oil field gets a customed-tailored MER. In Texas, only the richest fields are so closely watched by the regulators. Production in other Texas fields is limited in another way, by uniform rules on the spacing of oil wells. These rules specify the maximum allowable daily production for wells drilled to a given depth and draining a given amount of acreage.

Every One Is Unique

The ostensible purpose of both MER and the depth-acreage rule is the prevention of physical waste. The regulators dedicate themselves to this task with considerable zeal. But their tools are inadequate to their stated objectives. The depth-acreage provision is obviously deficient. Petroleum engineers never tire of pointing out that every oil reservoir is unique. The rules that would minimize so-called physical damage to one might be completely inappropriate to another. In applying a uniform standard to so many disparate reservoirs, the state in effect guarantees that some will be producing beyond the point of alleged damage, while others will be underproducing. Yet officials of the Texas Railroad Commission, which regulates oil production, have an annoying habit of stating publicly that all fields are now producing at their MER.

The concept of MER itself is singularly fuzzy. Some oil experts don't believe that there is any such thing as a production rate that will maximize total recovery. They maintain that the amount of oil eventually recovered from a given reservoir is usually independent of the rate of current production. Says Folkert Brons, a professor at the University of Texas and one of the world's leading authorities on reservoir engineering:

> The concept of MER has been pushed by geologists and economists. They are fine people. But they don't have the faintest idea of how an oil reservoir behaves. Most reservoirs cannot be dam-

aged by overproduction. Some may be damaged, but, in my experience cases of permanent damage are extremely rare. And a few reservoirs are actually helped—that is, the faster you produce, the greater the ultimate recovery.

Brons agrees that if you "overproduce" the giant East Texas field, where oil is displaced primarily by the encroachment of water from below the oil sands, you will unquestionably reduce water pressure and increase the amount of gas saturation. As water pressure declines, it becomes difficult to lift oil.

But the problem can be solved. Says Brons:

> In time, water will seep back into the formation to rebuild the pressure. And if you want to accelerate the process, you can always force water back into the ground. Once the pressure is rebuilt, the increased amount of gas saturation may actually strengthen the expulsive force of the original water drive, and this could mean a greater recovery of both oil and gas than would otherwise have been possible.

The Problem Is Economic

But the principal objection to MER has to do more with economics than with geology. Strict application of the MER concept requires that current production be curtailed whenever there is a threat to future production. Yet the state authorities make no attempt to compare the value of future output to the value of current output.

This omission must strike many businessmen as somewhat bizarre. When a company decides to invest in, say, a piece of machinery, it normally differentiates between the present and the future value of the money being invested. It is a truism that a dollar earned next year is worth less than a dollar in hand, and so the net profits expected from any investment must be discounted to their present value. One might, for example, multiply net profits expected in any future year by a discount factor based upon the "opportunity cost" of capital—that is, what could have been earned if the equipment had not been bought and the funds were used instead to buy financial assets like Treasury bills.

If the sum of the discounted net profits exceeds the current cost of buying the equipment, the investment is presumably worthwhile. If the discounted profits fall short of current costs, the investment should not be made—i.e., the businessman would be better off with those Treasury bills.

A similar present-value calculation is appropriate in deciding whether to pump oil now or in the future. If one expected oil prices to continue rising rapidly, then the value of oil in the ground would be rising with it, and rational businessmen would find it profitable to defer production as long as possible. But if price rises were not anticipated, it would pay to pump the oil now and use the proceeds from selling it to buy financial assets.

Betting on $22 Oil

If prospects did indeed warrant a sizable premium on future, as opposed to current, production, then the MER regulations might actually make economic sense. But it is hard to believe that the premium is really as high as our output restrictions imply. The price of oil is now about $10 a barrel. If the $10 were realized today and invested at 8 percent, it would amount to $21.59 in 1984. Those who want to defer production are, in effect, betting that the price of a barrel of oil ten years from now will be higher than that.

But it is most unlikely that oil prices will approach that level within the next decade. Indeed, they are more apt to fall. The rate of growth of demand for oil products is beginning to decline throughout the world, while the supply of oil and alternative fuels is expected to increase sharply. The United States has committed itself to an ambitious program to foster the development of substitutes for natural crude oil—nuclear power, shale oil, coal, synthetic oil (liquefied coal), and synthetic gas (gasified coal). The Federal Energy Office estimates that by 1985 the share of natural crude oil and natural gas in domestic energy production will decline dramatically as the United States turns increasingly to substitutes. Use of coal products, including synthetics, is expected to increase substantially. And nuclear power, hydropower, and geothermal and solar power will provide about 11 percent of all domestically produced energy in 1985 (versus less than 5 percent now).

The major oil-exporting countries may themselves contribute to the eventual collapse of today's oil-price boom. Their future behavior will be determined by the speed with which Western Europe and Japan—the chief oil importers—develop their own energy sources. It is technically possible, for example, to achieve full production of the North Sea oil fields by the end of this decade. According to Peter Odell, who is the director of the Economic Geography Institute in Rotterdam, full exploitation of the North Sea can provide Western Europe with more than half of its total oil and gas requirements by 1980. Currently, Western Europe produces only about a tenth of what it needs, relying on the Middle East for much of the rest.

As their chief customers become increasingly self-sufficient in energy, the oil-exporting countries will begin losing a good deal of their monopoly power and may be forced to hunt for new sales outlets. Instead of withholding their oil from the United States in an effort to raise prices, the exporting nations could easily reassume the role of zealous salesmen.

If the supply of imported oil and domestically produced oil substitutes increases substantially, the present value of any US oil that is still in the ground by the mid-1980s would be very much lower than most people now believe. From the viewpoint of the US oil producers, the failure to get this oil out as rapidly as possible in the 1970s would constitute a colossal economic waste.

If we adhere strictly to the concept of MER, we may avoid the waste of physical resources. It is true that if we exceed MER, we may end up damaging some usable resources. The damage might be catastrophic if we were without alternative sources of abundant energy. But our problem is centered on that "middle period," not in the post-1985 world—and after the middle period we will surely have abundant energy. Thus the sacrifice of some future oil recovery may not greatly matter.

There is something almost comical in the prospect of regulators battling to conserve oil as though it were an endangered species. We should not value oil for its own sake, as we value wildlife, but for its contribution to economic welfare. The United States has vast resources of natural crude oil, and the problem of "waste" is apt to be magnified by regulations that keep it in the ground when it could contribute to our welfare. Regulators who believe that they are preventing waste by keeping the oil in the ground have got things backward.

For oil, too, is an industry affected by the law of diminishing returns. The more we produce, the more difficult it becomes to extract oil and the higher the cost of production. (The average cost per barrel in the United States is now estimated at about $1.50, which, of course, is far below its price.) According to basic economic principles, we should continue producing oil until the cost of the last barrel is equal to the cost of the most economical substitute.

When might that be? Calculations made by William Nordhaus, a professor of economics at Yale University, suggest that, by the year 2020, it will not make sense to produce very much natural crude oil at all. By then we should be running the economy almost entirely on nuclear power, coal, shale, and synthetic oil, and on natural and synthetic gas, because *all* of these sources of energy will be cheaper than natural crude.

Officials of the state regulatory agencies tend to recoil at the notion that natural crude oil will ever be superseded. Their reaction is perfectly understandable. Having spent much of their lives working in the oil industry, they have developed an almost superstitious reverence for the stuff, and this influences their attitude toward waste. Says Arthur

Barbeck, chief engineer for the Texas Railroad Commission: "We will not allow any waste of oil at all. It would be foolish to produce to the point of sacrificing future oil recovery in order to guarantee that some fellow in Vermont or in Massachusetts does not have to stand in a line at his filling station."

This formulation of the case overlooks some options available to that fellow. If he has to stand in line whenever he needs gas, he may decide not to take a planned vacation trip. If enough people make such decisions, motel and resort business will fall off—and people will be thrown out of work. That really is waste. A fetishistic preoccupation with physical conservation victimizes everyone—not merely those fellows in Vermont and Massachusetts, but also those in Texas who will suffer from the effects of growing nationwide unemployment.

If major oil-producing states like Texas suspended MER and depth-acreage restrictions, production could rise rapidly. We might well see an expansion as great as the one that occurred during World War II when output increased from 1.4 billion barrels in 1941 to 1.7 billion barrels in 1945, or more than 20 percent.

But the period just after the war is somewhat more like the present, and the oil industry's performance then might afford a better guide to what is possible today. From 1946 to 1948, the industry enjoyed an unusually powerful sellers' market. Demand was buoyant, indeed, insatiable—there was a heating oil shortage in the Midwest in the winter of 1947 —and prices went all the way from $1.41 a barrel in 1946 to $2.60 in 1948. The price increase in constant dollars amounted to 52 percent.

The industry responded energetically to the higher prices. Output rose by almost 17 percent, from 1.7 billion barrels in 1946 to just over 2 billion barrels in 1948. Operators drilled new wells and increased the productivity of old wells by redrilling them. The number of new wells in use went from 15,851 to 22,585, up by more than 40 percent.

Average daily output per well rose from 11.3 to 12.8 barrels. By 1948, the expansion of supplies had stabilized prices.

Pouring Water on Troubled Oil

Can the industry equal or surpass this performance during the next few years? There are grounds for tempered optimism—especially if those state restrictions on output are lifted. (The boom after the war took place despite tough enforcement of MER regulations.)

In the 1940s there was no thought of new energy technologies coming on-stream. In fact, oil was *the* new technology. Operators then had no reason to worry that competition from oil substitutes would undermine the structure of future prices. Now they have plenty of such worries—and every reason to step up current output.

In addition, the technology of oil recovery was still in its infancy in the early postwar period. Nearly all recovery was primary—that is, either the oil flowed to the surface under the pressure of the reservoir's natural drive or, if reservoir pressure was weak, the oil was lifted by pumps. Since the 1940s the oil industry has perfected methods of secondary recovery and is now working on tertiary recovery.

These methods of recovering oil involve a number of artificial techniques for rebuilding reservoir pressure once it has been weakened or exhausted. Secondary methods include simple water injection—that is, flooding the area under the oil sands—hydraulic fracturing of the oil formation, gas injection and (when the unrecovered oil is highly viscous) heat injection. Tertiary methods generally involve one or more of the above, closely followed by the injection of surfactants—e.g., chemical detergents.

Right now primary recovery is gathering in only about 20 to 30 percent of the total oil in the ground. Secondary recovery can double that proportion. And tertiary methods can boost total recovery to 75 percent.

Flooding Becomes Attractive

It is true that even the cost of secondary recovery is still high and the cost of tertiary recovery is a lot higher. When oil sold for $2 to $3 a barrel operators found it economical to water flood only some fields. But when the price gets to $7 or $8 a barrel, water flooding becomes an attractive proposition for most oil men. That, at least, is the contention of Myron Dorfman, a petroleum engineer at the University of Texas who is also an independent oil operator.

After a field has been water flooded output will not rise immediately. It takes anywhere from six to eighteen months for the water to rebuild the original pressure. But after that time, production could increase quite substantially.

Henry Steele, a professor of economics at the University of Houston, has calculated that if prices settle at around $8 or $9 a barrel, a lot more production would come on. It would become profitable to use a great deal of secondary and even some tertiary recovery methods. Steele believes that US oil production could increase by over 6 percent within three years—from the present level of about 9 million barrels a day to nearly 15 million.

He admits that his estimates may be "on the high side." Suppose that we halve them—but add back some of the potential for increased production simply through more intensive application of primary methods. On these assumptions, output might easily increase by around 40 percent within three years. The United States would then be producing close to 13 million barrels a day. We would not be near self-sufficiency in oil, but our dependence on imports would be materially reduced.

This outlook would improve still further if the cost of tertiary recovery could be reduced. Many experiments with these methods are now under way in the United States, but the process is still not economic. The cost of the surfactants is still too high.

But it might be lowered rapidly. Marathon Oil has perfected a method of tertiary recovery, known as "Maraflood," that involves injecting chemical detergent into the oil formation after water flooding. It is estimated that about one barrel of surfactant is needed for every three to five barrels of oil eventually recovered. Says Edward Heidinger, chief economist for Marathon: "The cost of the surfactant is high because we now make it in small, customized batches. If demand increases—which we expect—we will consider building a large plant to mass-produce it. Costs could be substantially lowered."

Marathon seems to be confident that its process has a lot of potential. The company has been out combing the countryside, trying to buy up leases on "tired" and abandoned oil properties.

Turning On the Gas

In contemplating the possibility of expanded oil production, we should not forget that it would come with a major fringe benefit: production of natural gas would also increase. Gas is usually found along with oil, either entirely dissolved in the oil or in the form of a "cap" located above it.

But the bulk of gas produced in the United States—about two thirds—comes from the so-called nonassociated gas wells, i.e., wells that do not also produce oil. And the potential for expansion may be even greater for nonassociated gas wells than for oil wells. There is no MER for gas reservoirs, and no real need for secondary or tertiary recovery methods. Primary recovery gathers in about 85 percent of the total gas in a reservoir.

Some conservationists have been claiming that we must husband our gas, that supplies are not really so abundant. The gas companies themselves claim only 266 trillion cubic feet (TCF) of reserves, equal to about twelve years of current output. But figures published by the Potential Gas Committee, a research group composed largely of industry representatives, suggest a good deal more abundance. They show an additional 257 TCF lying around unexploited in existing gas fields—and in close proximity to existing pipelines.

There is no real mystery about the status of these additional supplies. Producers have been sitting around waiting for prices to rise before tapping that gas. And the industry's potential does not end there: undiscovered reserves are estimated at anywhere from 1,100 TCF to 2,100 TCF. All of which means that there is probably enough gas to last at least through the middle of the twenty-first century.

The Role of the FPC

The outlook for gas prices and production is materially different from the oil outlook. Once an oil producer surpasses his September 1972, output, he is permitted by the Cost of Living Council to sell all "new" oil in a free market. But the price of gas sold in interstate commerce is still regulated by the Federal Power Commission, which is gradually allowing increases. The major oil-price increases are probably behind us, but the biggest gas-price increases may still be ahead.

Thus oil and gas producers will make quite different calculations about the advantages of keeping their products in the ground. Oil men are likely to assume that the current market value of what they have down there is greater than the present value of anything they pump out several years hence. But a gas producer might reasonably come to the opposite conclusion—and find it profitable to withhold his supplies.

Does this mean that we should end all federal regulation of gas immediately? Probably not. If we did so, prices would, of course, shoot up, and producers might then decide that the peak had arrived. But natural gas is a substitute for oil. And if gas prices did increase precipitously, then the demand for oil would resume its rise, and we might find oil men holding back production again.

Fixing the Price Incentive

How might we maximize production of both oil and gas? Paul Davidson, a professor of economics at Rutgers University, believes that the price of gas must rise but that the FPC has recently been letting it rise too rapidly. He would confine annual increases to a rate just less than the opportunity cost of capital.

In other words, if producers can invest the proceeds of interstate gas sales in short-term financial assets yielding 8 percent, Davidson would have the commission limit annual price increases to 7 percent. This would place a slight premium on current rather than future production, i.e., gas in the ground would not appreciate as rapidly as money invested in financial assets. Operators would be encouraged to begin drilling for that additional 257 trillion cubic feet.

Given the appropriate price incentive, gas production could soar. It has been estimated that by 1977 production might easily rise by about one third—from 22.3 TCF a year to 30 TCF. Combine that gain with the prospective 40 percent increase in crude oil output and you get an overall gas and oil increase of approximately 35 percent.

These calculations are admittedly rough, but there is no doubt at all that the potential for a rapid rise in output is there. All that is required is that we know how to take advantage of it. Energy companies must become convinced that it makes sense to produce now rather than to wait. If we revise those state and federal policies that encourage the companies to think otherwise, huge increases in energy supplies will follow within the next few years.

THE LOOMING OIL BATTLE OFF THE EAST COAST [4]

For weeks the President's Council on Environmental Quality [CEQ] has been making news with its 331-page report on oil and gas exploration in the Atlantic and the Gulf of Alaska. So when the official version finally appeared last Thursday [April 25, 1974], it got little fanfare. But it certainly did not escape notice, for the CEQ report sets the stage for the government's position on drilling on the outer continental shelf—in the Atlantic as well as other wildcat waters. And this ranks as the hottest battle between the oil industry and environmentalists since the Santa Barbara spill. At stake are enormous petroleum resources that offer the United States its best hope of achieving energy independence in the 1980s.

In general, the report recommends that the government proceed with leasing, at least in the Atlantic, "under carefully stipulated and controlled conditions." Happily, the most promising sites tend to be in areas that the CEQ regards as having the least environmental risk. But the CEQ report contains a host of opinions, recommendations, and warnings that are bound to keep the oil industry at loggerheads with an aroused public for some time. And its cautious wording indicates that exploration in rougher waters—where oil men expect to find much more petroleum than in the Atlantic—will meet even stiffer resistance.

Promising Sites

Though the Atlantic is no Persian Gulf, geologists say there are three main formations in the East Coast's outer continental shelf that are likely sources of oil and gas: Georges Bank, a big fishing area about 130 miles off New England; Baltimore Canyon, a trough that runs from Long

[4] From article in *Business Week*. p 80-4. Ap. 27, '74. Reprinted from the April 27, 1974 issue of *Business Week* by special permission. © 1974 by McGraw-Hill, Inc.

Island all the way south to Virginia and lies about 70 miles offshore; and an area 5 miles to 25 miles wide in the Southeast Georgia Embayment.

According to the CEQ, eastern Georges Bank offers the least environmental risk. After that comes the southern tip of Baltimore Canyon, western Georges Bank, central Baltimore Canyon, northern Baltimore Canyon, and Georgia Embayment.

The council bases this ranking primarily on the probabilities that unrestrained oil spills would reach either biologically productive wetlands and estuaries or recreational beaches. Spills in eastern Georges Bank, for example, would reach shore only 15 percent to 20 percent of the time in the spring and practically never in winter, because the winds and currents in that area are usually offshore. The probability jumps as high as 50 percent in the western part of the bank, says the CEQ, but "oil on the rocky shores of New England would, in general, be less damaging than in the salt marshes and wetlands of the middle and south Atlantic."

That generalization irks Representative Gerry E. Studds (Democrat, Mass.), whose district includes Nantucket, Martha's Vineyard, and Cape Cod. "We've got sand beaches on Cape Cod," he snaps, "not rocky shores." Studds also complains that the CEQ did not devote sufficient attention to the possible impact on fisheries, and he doubts the council's prediction that offshore production on Georges Bank would create nineteen thousand new jobs. The CEQ projects that New England could be getting 30 percent of its crude oil and 70 percent of its natural gas from the bank by 1985.

Weighing the Risks

An even greater economic boost might come from drilling in the Baltimore Canyon, though the impact would be less noticeable because of the large population in the mid-Atlantic states. The CEQ estimates that offshore production might provide 10 percent of the region's oil and gas by 1985 and create thirty thousand new jobs.

But the environmental dangers are greater, too. In the central and southern parts of the canyon, where most projected drilling sites are 50 to 75 miles offshore, the risks are similar to those off New England. But in the northern section, the probability figures soar. An uncontained oil spill 25 miles off Long Island, for example, has a 75 percent chance of reaching shore; if the spill occurs 10 miles out, the probability approaches 100 percent. Northern New Jersey, site of some of the nation's most heavily used beaches, would also be highly vulnerable.

The CEQ is even tougher on the Georgia Embayment. The report includes such a strong statement on the environmental dangers there that it amounts to a recommendation against drilling at all. Potential sites are much closer to land, notes the report; hurricanes sometimes whip up 90-foot waves; the shore is lined with both important ecosystems and beautiful beaches. Still, drilling could provide 15 percent of the region's oil and gas needs.

None of the Atlantic drilling sites is as bad as the nine promising sites in the Gulf of Alaska, however. The gulf is prone to earthquakes, hurricanes are common, and sea conditions are sometimes very severe. On the other hand, few people would be affected by an oil spill in this sparsely populated area.

Secondary Effects

But spills are not the only worry. Environmentalists also fear the onshore impact of support facilities, such as service centers, storage tanks, and new refineries. "We are particularly worried about Cape Cod," says Mike Ventresca, executive director of the Massachusetts Forest & Park Association. "There is a lot of land there that oil companies might want to use." To prevent abuses, the CEQ urges states to strengthen their coastal zoning and other land use programs.

Some critics are more worried about the problem of rapid growth. Malcolm F. Baldwin, director of the Washington-based Environmental Impact Assessment Project, points to the North Sea to illustrate what can happen when big strikes are made offshore. "In Aberdeen, Scotland, a university town of 180,000, speculation and housing demand have caused real estate prices to skyrocket," sighs Baldwin. "And increased harbor fees have forced the entire whitefish fleet to flee."

Others complain that the CEQ data—much of which originated within the industry—were inadequate. "Given the nature of the estimates," complains Barbara Heller, of the Environmental Policy Center, "I would seriously question the assumption that New England could obtain 30 percent of its crude and 70 percent of its gas from Georges Bank."

CEQ Chairman Russell Peterson bridles at any suggestion that the report is too soft on the oil industry. "We do not believe that any economic development should run roughshod over the environment," he replies. Peterson points out that the CEQ report is not an environmental impact statement, and he urges the Interior Department, which must prepare such a statement before it can proceed with leasing, to give highest priority to the low-risk areas cited in the report.

Oil Men's View

Because the CEQ report was more for drilling than against it, oil men were pleased—those, at least, who were willing to talk about it. "I think people are worrying beyond reason about the danger of pollution from offshore drilling in the Atlantic," says Owen D. Thomas, worldwide exploration manager for Phillips Petroleum Company. "We've been exploring in the North Sea since 1969 without incident, and we've had some terrible storms." Adds Joseph O. Carter, vice president of exploration for Gulf Oil United

States: "I can't say that we won't have a spill, but I can say that our record offshore is pretty good."

Despite a few horrendous incidents, in fact, the overall statistics are impressive. Though oceanologists say oil pollution on the high seas is getting worse, studies show that only about 2 percent of it comes from offshore production. Half of the remainder comes from ships, mainly tankers, and the other half is waste oil from onshore sources.

And there are signs that things are improving. Government figures show that from 1964 through 1972, oil companies spilled 0.011 percent of the 2.6 billion barrels of oil and natural gas liquids they produced in federal waters off Texas and Louisiana. In 1971, two years after the Santa Barbara catastrophe, the figure dropped to 0.0003 percent. In 1972, it was a mere 0.00004 percent.

Moreover, oil men argue that offshore production is the most environmentally sound way of increasing domestic energy supplies. They point out that a lot of natural gas probably lies beneath the outer continental shelf, and gas is by far the cleanest fossil fuel. Trying to make up the nation's petroleum deficit with coal will require massive strip mining, they add. And more offshore oil production is preferable to more tankers spilling a trail of imported oil along our coastal waters. To get the industry's message across, the American Petroleum Institute [API] is organizing a speechmaking tour for the East Coast.

Unwinding Red Tape

If this tactic can prevent court delays, the first lease sales for tracts in the Atlantic Ocean could come next year [in 1975]. In a novel move, the Interior Department has already asked the industry to recommend areas on the outer continental shelf (in the Atlantic and elsewhere) that should be put up for lease. It has also asked environmentalists to rank problems in the various areas. Responses must be in by May 1 [1974]. Advance seismic work will then be possible.

The Bureau of Land Management will next prepare an impact statement for the entire Atlantic leasing program, and that, plus presale research, could take as much as a year. The sale would probably take place late in 1975, and the first well could be sunk in early 1976. Assuming immediate success, production could start in the Atlantic by 1979.

But environmental precautions are not the only obstacle that the bureau must overcome. Five years ago, Maine and twelve of the original thirteen colonies filed a suit claiming that they, not the federal government, own the resources under the Atlantic's outer continental shelf. The states, realizing that practically none of the oil and gas in the Atlantic lies within their 3-mile limits, argued that the charters granted them by King George III were not relinquished when the Union was formed. All testimony is now in the hands of a special Supreme Court master, and the government hopes the case will be settled by the middle of next year. If not, the Atlantic leasing program will grind to a halt.

Offshore Reserves

All the infighting and environmental opposition might give the impression that the Atlantic will be a veritable wellspring of oil and gas. At one time oil experts thought it would be. In 1968, the United States Geological Survey [USGS] estimated that as much as 42 billion barrels of oil and 211 trillion cubic feet of gas would be found in the near Atlantic. But a great deal of seismic data has accumulated since then, and the USGS recently scaled its estimates sharply downward as a result. The latest figures: 10 billion to 20 billion barrels of oil, and 55 trillion to 110 trillion cubic feet of gas. "These are highly speculative numbers," points out one USGS official. "Additional information may well warrant substantial modifications."

Oil men now generally believe that the Atlantic Ocean will prove to be a lesser oil province than the Gulf of Mexico—even though much of the oil and gas in the gulf has already been produced. "There have been a lot of statements about the millions of barrels of oil that exist in the Atlantic," says Edward R. Prince, Jr., chairman and president of Digicon, Inc., which has surveyed various Atlantic areas each year since 1969 under contract to various consortiums of oil companies. "But about all you can say definitely is that there are some interesting possibilities. You won't know until you drill some holes."

So far, drilling in adjacent areas has been disappointing. There is not a single commercial well on the East Coast all the way from Florida to Canada. The only drilling in the North Atlantic itself has been off Nova Scotia, where the majors sank 30 dry holes before hitting oil off Sable Island, 100 miles offshore. And they still have not found a commercial field after 70 attempts.

Partly because Georges Bank is closest to Sable Island, oil men generally believe this is the most promising formation in US Atlantic waters. Next comes Baltimore Canyon, then the Georgia Embayment. Because this is also the order preferred by the CEQ, most oil companies will probably find little to complain about when they issue their formal responses to the CEQ report. Most but not all.

"We think Baltimore Canyon has the best overall potential for drilling—and not just from the standpoint of total recoverable barrels," says J. R. Jackson, an executive in Exxon USA's exploration department. "It's hard to say whether there is more oil at Georges Bank or Baltimore Canyon, so why start drilling at the area that is farthest out and has the roughest weather?" Jackson also disagrees with the CEQ's judgment that the Georgia Embayment is the worst of the three areas. "They studied a site 5 miles offshore," he explains. "But we think the oil and gas there is 50 to 100 miles out."

Elsewhere on the outer continental shelf, the environmental considerations conflict head-on with the generally accepted geological potential. The USGS estimates, for example, that the Gulf of Alaska, which the CEQ judges to be a dangerous place for drilling, contains twice as much oil and three times as much gas as the Atlantic. Other offshore Alaska areas look even more promising, notably the Beaufort Sea and Bristol Bay. So does offshore southern California, a deepwater, seismically active zone scheduled for leasing about a year from now.

Self-Sufficiency

Can these offshore fields make the United States energy self-sufficient? "The outer continental shelf is a key part of Project Independence," declares John Sprague, chief of [the Department of the] Interior's Marine Minerals Division. "It's by far the largest source of untapped hydrocarbons." Adds William A. Vogely, acting deputy assistant secretary for energy and minerals: "We need very aggressive leasing of the outer continental shelf." [The Department of the] Interior hopes to have a revised five-year leasing schedule out by late summer [1974].

The Federal Energy Office is also eager to see what may lie beneath the nation's coastal waters, but John C. Sawhill, the new FEO administrator [who has since resigned], is less frantic about it. "I'd like to see drilling on the shelf," he says, "but we don't have to go hellbent."

Oil men, however, agree with the Interior [Department] spokesmen, and they are anxious to see the government open more of the outer continental shelf to their drilling bits. "Thus far the petroleum industry has been able to obtain leases on only about 3 percent of this country's continental margin," notes Wilson M. Laird, of the API. "Yet from this we are now getting 17.6 percent of our domestic oil production and 14.9 percent of our domestic natural gas production." Indeed, the oil companies consider the continental shelf the last hope for the United States. "There could be some big fields out there—some really big ones," says Gulf's Carter. "And God knows we need them now," adds Phillips' Thomas. "Except for onshore Alaska, offshore United States is our last great exploration frontier."

THE WORLDWIDE SEARCH FOR OIL [5]

"If I were a betting man, I'd wager anything that we will never find enough big new deposits of crude to declare independence from the Middle East," says the exploration vice president of a major US oil producer. "There are simply not enough new places in the world where just the right geological conditions exist to have produced major oil deposits."

"We know where roughly seventy of these spots are now," he continues. "And most of the rest, maybe thirty or so, are probably inaccessible to us—either politically or because of limitations in known drilling techniques."

Most geologists agree. The chance of finding many more recoverable deposits of 1 billion barrels of oil or more, they say, are remote. Nevertheless, the painful repercussions of last winter's [1973] Arab oil embargo, followed by a fourfold increase in the price of crude from OPEC countries to $11 a barrel, have fueled a search for oil that surpasses anything that the oil producers have attempted before.

"It is just wild," says Douglas G. Garrott, deputy exploration manager for Exxon Corporation, the world's largest oil company. More than 2,600 onshore and offshore drilling rigs were operating worldwide at the end of 1974, a 20 percent jump from just a year earlier. By the end of 1975, the number of rigs engaged in the worldwide search for oil could approach 3,000.

The intensified exploration is producing results. Just since January 1 [1975], for example, the Indian Oil & Gas Commission announced that it has hit oil in a third ex-

[5] From article in *Business Week*. p 38-44. F. 3, '75. Reprinted from the February 3, 1975 issue of *Business Week* by special permission. © 1975 by McGraw-Hill, Inc.

ploratory well, some 80 miles northwest of Bombay, that promises to produce up to 80 million barrels of oil a year by 1980. Amoco Norway, working in collaboration with the Norwegian Oil Consortium, Amerada Petroleum Company of Norway, and Texas Eastern Norway, Inc., has had a 2,695-barrel-per-day oil strike in the Norwegian sector of the North Sea. And below the Sadlerochit Formation, the main oil and gas deposit in the Prudhoe Bay area, a team of Atlantic Richfield and Exxon drillers think they may have located a big new oil formation, not previously figured in North America's 55.6-billion-barrel reserves.

Such discoveries would have to be repeated countless times to match the oil reserves of the Middle East, now estimated at about 367 billion barrels. Meanwhile, several serious problems could put a brake on the exploration drive:

☐ The cost of everything connected with oil exploration is soaring. "We've already seen a slowdown due to the oil industry's capital limitations, and we'll see more of it before the end of the year," predicts Garrott.

☐ Scores of foreign companies have invaded what was just a few years ago a predominantly US business, and this has raised the threat of temporary overcapacity in parts of the exploration business. Companies that have made commitments to purchase equipment but have nowhere to sell their services could find the going very difficult by the end of the year [1975].

☐ The political dangers involved in exploration in some areas are getting more intense. Oil companies will take calculated risks. "After all, that is what exploration is all about," says Wilfred P. Schmoe, Continental Oil Company's production manager for the Eastern Hemisphere. On the other hand, an increasing number of governments are "making it all but impossible to set up any kind of mutually profitable production agreements." Says Schmoe: "It doesn't make sense to put in massive exploration money when you know you don't have a prayer of getting it out."

Whatever new production comes from the oil companies'

intensified worldwide search, it is unlikely to bring down petroleum prices. Today's costly exploration efforts are predicated on oil prices going no lower than the current world market price of about $11 a barrel, and possibly rising still further. That is one reason why the exploration business is still a good gamble, despite higher costs, sharpened competition, and political threats. No one knows that better than the companies that share in the $10-billion-a-year oil exploration industry.

As the Big Companies See It

The oil companies' role in oil exploration is basically that of overall boss of the operation. Oil companies call the shots on where oil drilling will take place. But except for a few of the biggest of them—such as Exxon, British Petroleum, and Royal Dutch/Shell—few own or operate much of their own drilling equipment. Instead, they lease such equipment—most of it two or three years in advance and some of it even before it is built—from drilling contractors who charge them a daily rental that ranges up to $70,000 a day in the case of a new semisubmersible offshore drilling vessel.

Before an oil company decides where to drill, it has already spent an average $5 million on preliminary geophysical surveys and analysis in a 10,000-square-mile area. Onshore, this technical assessment is generally fairly inexpensive. But offshore, where most of the remaining unproved major oil reserves are thought to exist, it is far more costly. Sophisticated magnetic, gravimetric, and seismic surveys must be used, calling for ships, helicopters, and underwater equipment.

With the best geological data in hand, the oil company still has no proof that it has found oil. Only a 9-inch-diameter hole, drilled sometimes to 30,000 feet below the earth's surface, will confirm the presence of oil. And offshore, the cost of drilling an exploratory well can exceed the $1 million cost of drilling one on land by 300 percent to 1,000 percent.

Before a company decides to make this investment, it considers two other things: the economics of oil production in the area and the possible political ramifications. "Parts of the South China Sea, as well as both onshore and offshore areas around Indonesia, have long attracted oil explorers," says C. Howard Hardesty, Jr., Conoco's Eastern Hemisphere Petroleum Division president. "Oddly enough, the political climate there is far more receptive to profitable oil production than in many other areas right now."

Nevertheless, despite long effort, few large oil fields have been found in Southeast Asia. So, despite the fact that exploration money continues to flow into the area, it is moving in at a fairly controlled rate.

By comparison, says Exxon's Garrott, "the possibility of finding big new oil deposits in the North Sea looks much better." This is the hottest area of oil exploration right now. There are 45 rigs working in the area, and 10 strikes, each promising more than 1 billion barrels of oil, have been made in the past five years. A huge market lies close by. But winter drilling in the North Sea can cost up to $7 million per well. Weeks of work are lost during the bad winter weather. Drillers must battle 90 foot waves and winds of up to 120 miles per hour. The logistics of supplying crews in the North Sea are extremely difficult, and operating a workboat can cost up to $3,500 a day. On top of this, there is now the threat of what oil companies consider "an impossible" British Labour government participation and taxation plan for oil found in its North Sea sector, and an almost equally difficult situation in the Norwegian North Sea sector.

All these questions get long consideration as an oil company sets up its exploration budget. And the question of competition is involved, too. Shell Oil Company, the US company 65-percent-owned by Royal Dutch/Shell, for example, did no exploration outside the United States and Canada until 1970 but now is committing 25 percent of its annual exploration funds to such places as the Amazon jungle, the west coast of Africa, and the waters off South Vietnam.

Companies confess to the problems and costs involved. "It's as expensive drilling in the jungles of Peru as it is on the North Slope," says Gerry A. Burton, vice president of Shell's International Ventures. "It takes us sixty days to get supplies to our drilling crews in the Amazon jungle, and it's 600 miles from our rig off Vietnam to our base in Singapore." Shell was late in getting into the worldwide search for oil and knew it could not compete in more accessible drilling sites against companies that had worked in those areas for years.

The Drilling Contractors

Once an oil company decides where to drill for oil and buys a drilling license, the drilling contractor takes over. The contractor owns the rigs and all the equipment—derrick, drilling pipe, and pumps. He hires the drilling crews and other support personnel, which range from food caterers to drilling-mud suppliers and "loggers" who continually chart the hydrocarbon content of material forced up from the bottom of a well.

Two years ago, the total annual revenues of the world's drilling contractors stood at about $2.5 billion. Today [in 1975] they are more than $4 billion, and there is no sign that they have reached a peak. Such a surge in revenues is one reason why parts of Texas are still having boom times while much of the rest of the nation is mired in recession.

Part of the increase is due to rising labor and equipment costs, which the contractors pass along in increased rates. Today, a new offshore jack-up drilling vessel for use in relatively calm water up to 300 feet deep costs $25 million, compared with only $13 million in 1973. A semisubmersible vessel for use in the North Sea now carries a price tag of $45 million, almost double what it was two years ago. A dynamically positioned drilling ship of the type being used off North Africa now goes for up to $60 million. And even a land rig that can drill to 18,000 feet costs $4 million now versus $2.5 million in 1972.

But the contractors' increased equipment and labor costs are not the only reason why the rates that they charge oil companies have doubled in the past few years. With demand for rigs running high, contractors can wring out more profit for their services—and they do not hesitate to do it. "Drilling has been a depressed industry since the late 1950s, but now it is suddenly very healthy," explains William D. Kent, president of Reading & Bates Drilling Company, a subsidiary of Reading & Bates Offshore Drilling Company. "A few years ago, drilling contractors did not have the cash flow to buy more rigs. With our return on invested capital down at 7 percent, the banks wouldn't even look at us. Now, with our returns up to 15 percent, we're one of Chase's biggest borrowers."

Two years ago, Reading & Bates had 3 land rigs and 14 offshore rigs for lease; today it has 9 land rigs and 20 offshore rigs, the latest of which, a semisubmersible, was commissioned just last month. Similarly, Sedco, Inc., Ocean Drilling & Exploration Company, and Offshore Company—the three largest US drilling contractors—are adding new equipment at a furious pace. And they are signing contracts leasing the equipment to oil companies long before the rigs are delivered.

"The nut of our current profitability," admits Kent, "is that the cost of drilling equipment we'll get this year is based on 1973 costs, because that was when we ordered it. But we are leasing that drilling equipment to the oil companies at 1975 prices, because they represent what it would cost us to replace it."

Operating costs, according to Kent, are also up, largely because of a critical shortage of drilling crews. The monthly pay of an average worker on an offshore rig has risen 50 percent in the past three years, and it is still rising. A driller on an offshore rig in Southeast Asia, for example, now makes about $2,400 a month; in the North Sea he commands $2,500 a month.

US drilling contractors blame much of this cost increase on the Norwegian shipowners who, along with the Japanese, have been pushing into what traditionally has been a business dominated by US-based companies. "On wages and benefits, the Americans used to stick together," says an executive of a major drilling contractor based in Britain. "But the Norwegians started building rigs and stole people away. Now the US control of the drilling contractor industry is down from 95 percent to roughly 70 percent, and we have to meet the going labor rates to keep our rigs working."

New Foreign Competition

Recent new foreign entries into the drilling contractor business include Smedvig, Waage, Wilhelmsen, the Aker Group of shipbuilders, Ditlev Simonsen, and Ugland Shipping Company.

"The Norwegians are going to be a major force in the contract drilling business," predicts Kent of Reading & Bates. "Most of them are wealthy shipbuilders, and they are ordering rigs like they are going out of style. It's because of the tax structure in Norway. If their profits from shipping are not reinvested in new marine equipment, the taxes they would have to pay on them would be confiscatory. They have huge profits from shipping, and since there is a surplus of tankers already, they are investing those funds in offshore oil equipment to protect them from taxation." The result is a swell of orders for new rigs that could create a surplus in the North Sea this year and in Southeast Asia next year.

"There is a growing, but we think temporary, softness in demand for the heavy-duty, severe-weather semisubmersible rigs for the North Sea," admits Michael R. Naess, an executive vice president of Zapata Corporation, a US drilling contractor that recently entered into a deal with Norway's Ugland Shipping Company. Under terms of the deal, Ugland will put up all the capital for buying new rigs and will lease them to the joint venture at cost plus financing charges. Zapata, in return for supplying drilling expertise and personnel to the joint venture, will get 70 percent of the profits; Ugland will take the other 30 percent.

Other companies have also entered into joint ventures. And this is adding to oversupply. "The Norwegian sector has not created all the drilling demand that was anticipated," Naess says, "and development has slowed down even in the British sector. People are concerned about the Labour government's attitude on ownership of the oil and on taxes."

Zapata and some of the big international oil companies say that they have had offers lately from nervous Norwegian shipowners who want to sell rigs that they have ordered but on which they have not yet taken delivery. "At the rate that some companies have been ordering rigs, there is sure to be an oversupply by late next year or in 1977," agrees Exxon's Garrott. "Some people are going to go broke."

At Sedco, largest of the oil-drilling contractors, with drilling revenues last year of $116 million, Spencer Taylor, president of the Drilling Division, agrees. "There are going to be some rigs in the North Sea that will have to be shut down for some period," says Taylor. "Those that do not have drilling contracts yet are going to have a pretty rough time."

If costs and government policies are not enough to cause oil companies to slow down their drilling plans in the North Sea, the shortage of trained drilling crews will brake the wild rush. Sedco and other big contract drilling companies have set up schools to train more drillers. Sedco now has three such schools in operation (in Scotland, Algeria, and Iran) at an annual cost of $1 million. But there still is a serious shortage of qualified people.

This, too, could explain the recent drive on the part of many foreigners to link up with American partners in joint drilling company ventures. They know that without expert drillers, they cannot get contracts for their drilling equipment. And beyond a certain point, the available pool of trained manpower cannot be stretched.

The Boom in Offshore Vessels

If drilling contractors foresee the forecast of an oversupply of drilling vessels, the makers of those vessels have yet to see a slackening of orders. No segment of the drilling industry has grown faster than the offshore drilling vessel construction business in the last few years. Led by US companies such as Marathon Manufacturing Company, Bethlehem Shipbuilding, Avondale Shipyards (a subsidiary of Ogden Corporation), and Levingston Shipbuilding Company, and joined by foreign companies such as Aker (a group of Norwegian companies), Rauma-Repola (of Finland), Robin Loh (a shipbuilder in Singapore), Mitsui, Mitsubishi, and France's CFEM, the industry collectively has built capacity to turn out fifty vessels a year—nearly double its capacity just three years ago. Its annual business volume now is $1 billion, a fourfold increase since 1972. And construction facilities are stretched to the breaking point.

Marathon, the largest company, for example, now has a $390 million backlog of drilling vessels on order, running to 1978. And Avondale, the only major US offshore drilling vessel maker without a shipyard abroad, has a $140 million backlog.

Unlike drilling contractors, whose profits have soared as a result of the sudden increase in the world search for oil, most offshore drilling vessel makers have not yet seen a similar profit boom. They admit that when the offshore boom started in the early 1970s, many of them underbid on new vessels because of a bad underestimate of the time and difficulty involved in construction. And some got caught with fixed-price contracts, with no escalator clauses.

"So far we've lost money on this drilling boom," admits Gene M. Woodfin, chairman and chief executive officer of Marathon. Suppliers continue to increase his company's costs to double what they were three years ago. As a result, two Marathon yards, at Brownsville, Texas, and Clydebank, Scotland, were still in the red last year, and the company's Rig Division barely broke even on sales of $110 million. But things are now looking up.

"If Exxon had lost proportionately what we lost in 1973—$19 million on full company sales of $252 million—it would have gone out of business," says Woodfin. "It was a miracle for us to survive. But now we can see the end. And we don't intend to subsidize the drilling industry again."

M. Lee Rice, president of Ogden Transportation Corporation, the subsidiary of Ogden Corporation that operates the Avondale Shipyards in New Orleans, also admits to bad mistakes when his company reentered the offshore drilling vessel business back in 1971. "The first two vessels we contracted to build were semisubmersibles," he says. "Together, they lost us about $300,000."

Now, however, things are different. Ogden quotes contract prices, according to Rice, only on an escalating basis, pegged to the cost of its materials and labor. It has built another seven or eight—all of which were profitable. And it picks and chooses contracts that it feels fit its building techniques and capabilities. But it and other builders have faced tremendous cost increases because their material suppliers, primarily the steel companies, have pushed up prices faster even than the general inflation rate. "The cost of building rigs has doubled, with material cost increases providing the driving force," says Rice.

The Competition in Rigs

The largest maker of drilling rigs, the machinery used to drill oil wells, is the National Supply Division of Armco Steel Corporation. With the capacity to build about 50 drilling rigs a year, it splits with Lykes-Youngstown Corporation's Continental-Emsco Division and United States Steel Corporation a full 80 percent of a $600 million annual business—now about 200 rigs a year. About a quarter of these rigs are sold for offshore platform use; the rest are sold for land exploration.

Many of these same companies make drilling pipe—the 30-foot lengths of pipe that when cemented together in a drill hole form the well casing. Until the last couple of years, drill pipe was a low-profit item for steel producers, and last year, with a squeeze on steel capacity, many of them quit making pipe. One result: Pipe prices soared 150 percent in the United States last year. Now, with fatter margins available on drill pipe, other producers are getting into the business. French and German steel makers are now producing it, and Japanese mills are joining in, too. Three Japanese steel makers are offering drill pipe for sale in Britain. They have won a joint contract to supply the specialty steel that will bring oil out of the North Sea's biggest oil find: the Forties field, which contains an estimated 4.4 billion barrels of oil and which lies under 400 feet of water, 115 miles off Aberdeen, Scotland.

Likewise, Japan has contracts to provide about 18 percent of the first 37,000 tons of steel needed for the first two Forties oil production platforms. And it has been a big pipe supplier to the Alaskan North Slope.

While such new suppliers have helped reduce the materials shortages, they have done little to stem the upswing in prices, the rig makers complain. In addition, there is very little doubt that the Japanese, like the Norwegians before them, had to move their steel as quickly as possible into an even bigger profit area—building not only offshore drilling vessels but production platforms to recover whatever oil the fifty oil companies now drilling in Southeast Asia discover there.

"The collapse in the supertanker business has left Hitachi, Mitsui, and Sumitomo with a lot of steel on their hands, and you can build a helluva lot of drilling vessels or platforms with the steel it takes to build a supertanker," says George Morris, who heads Marathon's Singapore shipyard. "The Japanese have the basic steel-fabricating expertise, the technical skills and engineering staffs, and licensing arrangements with the leading US rig designers," he continues. And their yards are closer than European and US companies to the scene of operations.

Building the Platforms

If they move beyond the offshore drilling vessel business into making permanent offshore oil production platforms, the Japanese will find some fierce competition. Two US companies—Brown & Root, Inc., and J. Ray McDermott & Company—have long dominated this business, and they still control about two thirds of it.

In recent years, companies such as Uie of France, John Laing and Dorman Redpath & Long of Britain, and DeGroot of the Netherlands have made big investments in platform construction. The world platform business now runs around $1.5 billion per year. And it is just starting to feel the full force of the surge of orders flowing out of the accelerated search for oil. "It is a mystery to me why more shipyards around the world have not moved into the platform construction business," says John D. Dupy, a vice president of J. Ray McDermott. "It may be because a shipyard has a full line of craftsmen, and platforms require only a limited line, principally welders."

Profits, he says, are good and getting better. McDermott, which realized only $6.7 million on revenues of $238 million in the year ended March 31, 1972, showed profits of $34.7 million on a volume of $425 million in fiscal 1974. It is estimating sales in the $700 million range this year [in 1975].

Moreover, according to Dupy, the platform part of the oil exploration and production business is still moving overseas. Since 1972, Brown & Root has opened new yards to produce platforms in Brunei and Indonesia as well as in Nigg, Scotland, and has doubled its production capacity in Bahrain. J. Ray McDermott has new yards in Dubai on the Persian Gulf and in Indonesia as well as a new yard in Inverness, Scotland. As recently as 1971, the oil platform business was split just about 50–50 between US and foreign work. Now, only 30 percent of it is done in the United States.

US platform makers are not especially worried about this move, however. Nor are US builders of other oil-drilling equipment, because they see a huge potential for sales that is virtually untapped. That market is China and the officials of Peking's National Machinery Import & Export Corporation (Machimplex) who are already shopping for a long list of exploration gear, ranging from specialized drill bits to multimillion-dollar deepwater drilling rigs.

Nobody knows how much oil there is in China. Noted US geologist A. A. Meyerhoff is guessing at a minimum of 20 billion barrels of oil—or something less than in all of South America. Others think that is too conservative an estimate. Chinese government officials have apparently decided that oil is the answer to their country's deepening trade deficit, so they are prepared to spend to boost their crude production as rapidly as possible.

American companies that have received orders from Machimplex are usually reluctant to talk about them. The total so far is probably only about $20 million, including relatively small items, such as rock bits from Hughes Tool and Reed Tool, seismic survey equipment from Geospace Corporation, and various oil field equipment from Baker Oil Tools, Inc., of Los Angeles. The Chinese have placed far bigger orders with French, Danish, and Japanese companies worth more than $100 million.

"We are just nibbling at the edges of potential sales to China," admits J. Ray Pace, who runs Baker Trading Company, Baker's China-trade subsidiary. US regulations and restrictions on the export of some strategic items involved in oil exploration equipment have forced Peking to shop elsewhere, he and other Americans maintain.

The Surer World Markets

Enthusiastic as they are about the sales potential of oil exploration equipment and services to mainland China, US companies know they have a far surer market in such places as the Mediterranean, where drilling has picked up lately, and in the Persian Gulf, Africa, Venezuela, Southeast Asia, and the North Sea.

Longer-range, the action could start swinging back to the United States, especially if the government permits drilling on the outer continental shelf in the Atlantic. This will be far deeper and more difficult drilling than in the Gulf of Mexico. But Exxon already has a rig under construction in Japan that could work in Atlantic waters. It is being built at a cost of $50 million, and its first jobs will be exploratory drilling in 390 million acres of water down to 3,000 feet off the west coast of Africa, off Labrador, and off the West Coast of the United States. Amoco has a rig that can drill to 2,500 feet, and Gulf also claims to have equipment that can reach that depth.

Already, several of the thirteen US drilling rigs that were towed to foreign waters during 1971 and 1972—when environmental protests halted US offshore leasing—have been returned to the United States. And the White House move to place a $3 additional tariff on imported crude could cause other companies to reduce their exploration abroad.

A consensus among rig owners and oil companies, however, is that unless US tax laws are changed, some 75 to 80 other US-owned rigs will remain working outside the United States. The reason: Section 956 of the United States Internal Revenue Code would tax US owners of drilling rigs registered abroad at the rate of 48 percent of the adjusted value of the equipment at the moment the rig returns to US waters. "That's a steep price for returning a rig," says one oil company executive.

Nevertheless, the government is betting that if new lease sales are opened, US oil companies will find some way around this problem. One possibility would be to divert equipment that is under order for use abroad to new US sites. "An oil company plans its exploration budget four years ahead," says Exxon's Garrott. "But changes can be made. We are looking at a worldwide inflation rate right now that runs to 30 percent. With conditions like that, and with all the geopolitical problems, there's not much question that we'd sink more money in oil exploration at home if the prospects warrant it."

FAREWELL TO OIL? [6]

With their uncorrupted faith in the sublime dynamics of perfect competition, the editorialists of the *Economist* in London have been proclaiming a coming age of energy abundance in which oil producers will come hat in hand to sell their stuff at declining prices. According to this bastion of classic (i.e., nineteenth century) English liberalism, the present expectations of ever-higher oil and energy prices caused by increasing scarcity are based on the "third-rate political economy of linear projections," which take no account of the inevitable reactions to high prices and their long-run effect on the market.

This argument rests on a bit of hardcore economic theory: assume that a group of wicked monopolists were to corner the market for a product—say apples, a favorite in first-year textbooks—thus artificially raising their price to three times the premonopoly level. What happens next? The good student knows what he is expected to say: first, economy-minded consumers will start buying pears instead, thus depressing apple demand; second, fruit growers and nonmonopoly apple growers will plant many more apple trees, thus increasing industry supplies; and third, many other landowners will switch their lands to apple orchards,

[6] From article by Edward N. Luttwak, author and Visiting Professorial Lecturer at Johns Hopkins University. *Commentary*. 57:36-9. My. '74. Reprinted from *Commentary,* by permission; Copyright © 1974 by the American Jewish Committee.

thus increasing further the supply by expanding the size of the industry itself.

If oil were a product with the features of the apples and pears of the textbooks, the members of OPEC (the Organization of Petroleum Exporting Countries) would indeed be reduced to the ephemeral role of short-term speculators on a declining asset. Actually some of the oil, and specifically the 35 percent or so of total worldwide demand that is used for automotive fuel and feedstock—for which *only* oil will do—is not at all like apples or pears since there is *no* substitute, except for synthetics that cannot (at least not yet) compete with oil even at present prices. But the rest of worldwide oil demand, up to 65 percent of the total, is used as a source of general energy, and it can be replaced by other fuels. Thus, while the "specific" oil sector is governed by the monopoly OPEC supply and the unpredictable contingency of new oil finds (still more a matter of accidental discovery than scientific exploration), the rest of the oil market is subject to the normal rules of supply: if the prices go up so should the supplies, until prices decline once more.

But "general" energy supply is also subject to other constraints that never seem to find their way into the textbooks: long investment "lags" (up to nine years for a US nuclear power plant), extra-economic restrictions (primarily environmental), industrial bottlenecks, and not least, monopoly restraints—which are by no means confined to OPEC oil. To give one example, the export of Dutch natural gas—an important source of premium energy in northwest Europe—is controlled by a company with a long and unpronounceable Dutch name, which happens to be owned jointly by Shell and Exxon.

In the dim distant days of 1970 when the official or "posted" price of good quality crude oil from the Persian Gulf was still only $1.80 per barrel, and when the barrel actually sold for about $1.40 at the jetty, Arab and Iranian oil undercut all other energy sources the world over (though there were a few localized exceptions such as the artificially underpriced natural gas of the Southwest of the United States, Norwegian hydroelectric power in place, and Rhodesian coal). It was this unbeatable cheapness, and not the machinations of the oil companies, that brought about the relentless increase in the world's dependence on OPEC oil. The incentive to develop other energy supplies simply was not there. Instead there was a positive incentive to *abandon* existing, higher-cost energy sources, from the once great coal industry of France, Germany, and Belgium, to the windmills of Holland (maintenance was more costly than the diesel fuel needed for water pumps). It is this dependence that OPEC is now happily exploiting—and that the Arab oil producers have already exploited politically—and it is this same dependence that the *Economist* believes will be reversed, thus bringing about a collapse of the oil monopoly.

It is certainly true that at a "posted" price of almost $12 per barrel in the Persian Gulf, which currently results in real market prices of $8 or so, the incentive to develop alternative sources of energy must have increased by orders of magnitude. Further, in spite of the thick layers of taxation (outside the United States) that separate the cost of crude oil from the price of the refined products bought by the public, there has been a very sharp worldwide rise in product prices, and this ought to bring about some decline in demand. The two key factors that will determine the future of oil are thus the *flexibility* of supply on the one hand, and the *responsiveness* of demand to rising prices on the other: the price elasticities of supply and demand, to use the jargon of the economists.

On the supply side, almost every alternative energy industry with a real growth potential also seems to face intractable bottlenecks. In the case of nuclear power, a whole new industry must be created first in order to fabricate reactor components before much else can be done: until very recently very few reactors were being built and the reactor industry is naturally rather small. Even when the parts become available, nuclear reactors will not be built in large numbers very quickly since the concrete housings alone take several years to build. On top of all this, licensing procedures and the inevitable objections of the Naders of this world, both great and little, are apt to delay matters right from the earth-moving stage. At a rough guess, this means that it will be about fifteen years before the industry will be moving in high gear with lots of plants coming on stream and the reactor industry turning out parts for more at a rapid clip.

The coal industry has many of the same problems and also one of its own. In the mines of Belgium, Germany, and France, as well as in most British and quite a few American mines, cutting coal underground is perhaps the hardest and most dangerous job to be had outside the circus, combat reporting, and serving in Mr. Nixon's White House. It takes a special kind of man to go down into the mines, and it takes a special tradition to create such men, defiant of danger and discomfort—and submissive to poverty. There is not as much poverty around the coal fields nowadays and still less defiance. Experience has shown that once miners find other jobs when mines are shut down—as they have been all over Europe—the manpower is lost and the tradition is dissipated. It is then extraordinarily difficult to persuade normal workingmen, including ex-miners, to go underground again. In the late 1940s the British government introduced a special scheme to bring foreign workers into the mines with all sorts of inducements, but although there was much unemployment in the recruiting grounds of southern Europe, and real hunger too, only a small fraction of the tens of thousands brought over to work remained on the job. The broader coal seams of the United States, not to speak of strip mining, make the miner's job less impossible over here, but there is still a rather inelastic labor pool, and it will not be easy to convert out-of-work assembly line workers into coal miners.

Shale rock in the United States, the tar sands of Athabasca in Canada, the lignite of central Europe, and similar low-grade fossil fuels elsewhere have problems of their own too. In each case, vast quantities of rubble and rock have to be cut up, crushed, and processed in order to get a relatively small amount of fuel. All this crushing and grinding, cooking and refining needs a great deal of expensive machinery—which is in short supply since again the industry to make the industry must be created first. Moreover, it always seems to take rivers of water to get trickles of fuel, and we all know that water supply is already fairly tight where the shale and lignite are. The Canadians naturally have lots of water . . . , but a full-scale exploitation of the tar sands of Alberta would mean that half the province would have to be picked up and trucked away across the border, and the Canadians are on the restrictionist bandwagon just as much as Kuwait is.

As for the other elasticity, the responsiveness of demand to price increases, the instant energy experts created by the crisis hold two opinions. Some say that the world's energy demand is quite inelastic, their evidence being some such piece of high science as the statement that "people will drive no matter what"—I am quoting one of Capitol Hill's $400-a-day consultants. The others, spiritual brothers of the *Economist* crowd, are equally categorical in asserting that energy demand will surely decline. The data are on the side of the *Economist* crowd. Figures are already coming in from all over the world and, subject to seasonal uncertainties, it seems pretty obvious that Italians are no longer quite so keen to race each other on the *autostrada* for the fun of it, and that Indian peasants are running their diesel pumps slower, in the full knowledge that this will result in a hungrier harvest. . . .

But the energy industry worldwide also has a peculiarity that some (not all) of the enthusiastic price riggers in OPEC may have overlooked. Because of the very same rigidities that prevent the rapid and large-scale substitution of other energy sources for OPEC oil, most of the substitution that is and will be taking place is virtually irreversible. If the price of beef goes up, housewives can quickly buy chicken instead; when the price goes down again, they can return to beef just as quickly. In the energy industry, however, substitution will most often be a one-way switch. A nuclear power plant that is ordered in 1974 because it is more economical than an oil-fueled plant when imported oil is selling for $11 on the East Coast, will not come on stream until 1982 at the earliest. But once it *does* start working, it will remain in operation until the year 2000 even if the price of oil goes down again to precrisis prices, or even less.

Similarly, coal mines which are reopened on the basis of long-term supply contracts will be slow to come into production; it will take years to assemble the manpower and fit out the mines with modern cutting gear (the coal machinery industry is a major bottleneck in itself). But when the mines *do* come into operation, coal users and governments will no longer allow them to be driven out of business by imported oil, however cheap. As for shale oil, with its huge capital investments ($750 million for a modest 125,000-barrel-a-day plant), the ground rules are set by the high capital costs: they will be kept in operation no matter what once they are built.

The long-term implications of this unusual one-way substitution pattern are obvious. By allowing the oil price to go through the roof, OPEC has precipitated a worldwide rush to alternatives from which there is no return. The position of the oil monopolists is a happy one, but it is also fragile. The skillful monopolist in any industry must carefully weigh the extra gain of today's price increases against the future loss of market demand that such increases cause, and which cannot be recaptured. On paper, the $8 per barrel of the present [Persian] Gulf oil price looks like a reasonable bet for the producers. Unfortunately for OPEC, however, the manner in which the price was raised—abrupt, arbitrary, and seemingly open-ended—and more important, the context of the increase, the Arab embargoes, have defeated the monopoly strategy of optimizing price versus future demand. Governments worldwide are not acting as if the price of oil were $8 per barrel, but rather as if it were $16 or even $24 per barrel. Quite simply, no one wants to accept an indefinite dependence on OPEC oil; the Japanese, for example, have clearly decided to go nuclear as much as possible and as soon as possible, regardless of cost; more than the price, it is the insult that is propelling their program. The British and the French are accelerating the development of nuclear power, and North Sea oil and gas are now being developed at the fastest possible rate. The United States has launched its broad-front self-sufficiency program, and every other industrial country is moving at full speed in the same direction.

What this means is that OPEC is getting the financial rewards of $8 oil, but it is losing its markets as if the oil price were $16 to $24 per barrel. The robber barons of the Rockefeller era would not have made the same mistake. Thus, as the *Economist* predicts, from the 1990s onward OPEC oil may very well be harder and harder to sell to a world energy market dominated by new fuels that OPEC oil will no longer be able to undercut even if the price declines very sharply.

Why, then, did OPEC do what it did? The answer, of course, is that OPEC as a whole did nothing. First the Arab members imposed the output restrictions of the [1973] October war for noneconomic reasons of their own, and then the non-Arab OPEC members (Iran and Venezuela) exploited the induced shortages to raise prices. In the first instance, this obviously undermined the Arabs' political strategy. Offering or denying oil at $16 per barrel (some spot prices were even higher) does not generate the kind of political leverage that the manipulation of cheap oil supplies could have done. But actually the Iranian-Venezuelan strategy was a good deal more harmful than that of the Arabs. The Iranians and Venezuelans happen to be producing their oil so rapidly that their output will start to decline anyway in two decades or less. They imposed the high oil prices on the Arabs, thus maximizing their near-term revenues at the expense of post-1990 oil revenues when they themselves will be out of the market. This is a very cruel blow indeed to the Arabs who (with the exception of Algeria) are producing their oil at a much slower rate; unlike the Iranians and Venezuelans, they will still have plenty of oil reserves in the post-1990 period when oil will be harder and harder to sell.

Caught between the hammer of Iranian demands for ever-higher oil prices, and the anvil of radical Arab opinion which is always willing to suspect their motives, and which is utterly unfamiliar with twenty-year market projections, Yamani of Saudi Arabia and the other Arab oil ministers of the gulf were forced to go along with the catastrophic price increases in the full realization that this would enrich the Iranians—and undermine the long-term value of their own only asset. They know that Iran can industrialize while they cannot, since they have small populations and no other resources at all. When Yamani protests that he has been pressing other OPEC countries to *reduce* oil prices, he is telling the truth.

In September 1967, just after the six-day war, the British government published a white paper on fuel policy proclaiming the readiness of the United Kingdom to rely on Middle East oil indefinitely. In the paper it was acknowledged that political embargoes were even then in effect against Britain; it was also asserted that such embargoes would inevitably be short-lived and do no real damage; the cheapness of Arab oil justified continued reliance on it.

. . . [In 1974], in the wake of the 1973 war, no government would repeat these confident assertions. It is not merely that the embargoes this time were general rather than selective, nor that they were much more prolonged than in 1967. The real change is that the embargoes have now become instruments of a two-way political/economic squeeze. The new mood has been expressed by the *Petro-*

leum Press Service, a monthly oil journal which is generally pro-Arab and pro-oil company:

> The OPEC governments did more . . . [in 1973] than tear up contractual agreements, line their own pockets, and indulge in political blackmail. They forced upon the rest of the free world the urgent question whether it is tolerable that [the world's] energy supplies . . . should in future be subject to the will of a small group of exporting countries.

The question that the *Petroleum Press Service* so tentatively put has been answered decisively. It is clear that all governments which have any means whatsoever of reducing their dependence on imported oil, and in particular Arab oil, are now taking action almost regardless of cost. In December 1941 Japan went to war because the United States directly—through withholding gasoline—and indirectly—by deterring Japanese seizure of the oil-rich islands of Dutch Indonesia—was denying indispensable petroleum supplies to Japan. Today Japan is in no position to use military force directly to secure her oil supplies. The Europeans do have the military capability needed for the job, but they lack the political will to act even to protect their own economic survival. Both the Japanese and the Europeans, however, are and will be willing to make the economic sacrifices required to reduce their dependence on Middle East oil. To the extent that they succeed, Arabia can look forward to an economic future as desolate as its economic past. The ghost towns of the upper Amazon and the American West show how transitory an economic prosperity based on a single product can be.

NATURAL GAS SQUEEZE—HOW TIGHT WILL IT GET? [7]

Reprinted from *U.S. News & World Report.*

Signs are there is more natural gas available for US industry than Washington and the interstate pipeline industry have led people to believe.

In late January [1975], new supplies were suddenly forthcoming to keep factories open in several states, heading off layoffs of thousands of workers.

"Arm twisting" by the White House and Congress was judged to be the reason. Whatever the cause, these fast-moving developments served to ease critical shortages:

☐ United States Steel Corporation on January 21 said changes in a Federal Power Commission ruling would let it keep open three plants in the Pittsburgh area. Closing would have put 1,300 people out of work immediately and affected 90,000 tons of steel production.

☐ Factories in Danville, Virginia; Linden, Alabama, and Laurens, South Carolina, along with big agricultural-fertilizer plants in North Carolina and Delaware, won a reprieve when Transcontinental Gas Pipe Line Corporation found an additional 2 billion cubic feet of gas.

☐ The Federal Power Commission on January 21 ordered Tennessee Gas Pipeline Company to supply extra gas for thirty days to some customers who had faced cutoffs. Saved from shutdowns were factories in southwestern Virginia, three Aluminum Company of America plants in Tennessee and parts of the Atomic Energy Commission facility at Oak Ridge, Tennessee.

The Weather Factor

FPC officials were cautious about the outlook. They said that relatively mild weather had made more gas available and warned that a cold wave could reverse things. One

[7] From article in *U.S. News & World Report.* 78:47-8. F. 3, '75.

source said it was a case of "the squeaking wheels getting the gas."

Meanwhile, other industrial users of natural gas were seeking a bigger share of supplies. The volume of petitions for emergency relief to the FPC, which regulates the interstate gas industry, was running nearly double last winter's.

The FPC has started investigations into cutbacks by Tennessee Gas and Transcontinental. Hearings involving other interstate pipeline firms have also been scheduled.

Pipeline officials, for their part, said they were getting less gas from southwestern producers than they had been promised and had to cut back industrial customers to safeguard supplies for home heating. They cited problems that included side effects of Hurricane Camille, which swept through the Gulf of Mexico last September [1974].

Industrial users were converting to alternate fuels wherever possible, but natural gas is essential to some processes. The Radford, Virginia, plant of Lynchburg Foundry is an example. Its engineers have found no way to use other fuels in a shell-molding process. Without gas, the plant would have to close.

Officials from Danville, Virginia, a city of about 40,000, said as many as 10,000 would be jobless if plants there had to close because of gas cutbacks.

Holdbacks Denied

Left unanswered was the question: Are producers holding back gas in hopes of higher prices?

Pipeline officials told the hearing examiners they had no reason to suspect any such actions. Other gas industry officials said that the low level of federally controlled prices discourages new exploration.

Others are skeptical.

"If there is a satisfactory explanation of the shortage, it hasn't been put into a form which will satisfy the critics," said Lee C. White, a former FPC chairman who heads an energy study group for the Consumer Federation of America.

The Interior Department on January 23 ordered ten major oil companies to explain why some leases in the Gulf of Mexico are not being fully produced. If the reasons are not considered valid, the leases could be revoked.

Interior Secretary Rogers C. B. Morton [who is now Secretary of Commerce] said what might have been accepted business practice in the past would have to be reexamined in light of the vital need for gas. He said he would not allow companies to withhold production "in hopes of getting a higher price at a later time."

And the question is sure to be debated when Congress takes up President Ford's proposal to deregulate the price of newly found gas.

Haves and Have-nots

A striking paradox of the natural gas supply situation is its feast-or-famine aspect.

While plants in the East scramble for enough of the fuel, the state of Oklahoma is running national advertisements, inviting industries to relocate where there is plenty of natural gas.

What makes the difference? The answer is that gas sold in the state where it is produced is not subject to federal price controls. It now is selling for as high as $1.75 a thousand cubic feet, with abundant supplies reported in Texas, Oklahoma and Louisiana.

Natural gas which goes into the interstate pipeline network has a ceiling price of 51 cents a thousand cubic feet for recent discoveries. Much of the supply is under long-term contracts which range down to 20 cents or lower.

The result is predictable. While there has been a spurt in gas-drilling activity, less and less is leaving the state where it is found. Pipelines, which carry 90 percent of the supply for thirty-two states, have had to turn increasingly to offshore wells, which are committed to the interstate market under terms of their federal leases. Where pipelines got 75 percent of the gas onshore in 1970, they now get 33.

There also has been a relative falloff in offshore drilling, as rigs are concentrated in more lucrative areas. In 1971, 12.4 percent of all US gas exploration was offshore. In the first six months of 1974, the percentage dropped to 2.1.

Effect of Cutback

This trend accentuated an already severe drop in discovery of new reserves of natural gas. In 1967, a total of 21.1 trillion cubic feet was added to the nation's reserves, 70 percent of it earmarked for interstate pipelines. In 1973, discoveries were 6.5 trillion cubic feet. And just 17 percent of that was for interstate use.

"We simply cannot compete for new supplies of onshore gas," said Jim Thomas, an official of Texas Eastern Transmission Corporation, which has curtailed deliveries to both the East and West Coasts this winter.

But Houston Natural Gas Corporation, a Texas utility, ended its fiscal year with reserves 10 percent higher than at the close of the previous year. Said a spokesman: "We have no trouble in finding new supplies of gas if we are willing to pay the price."

With residential users getting top priority on available supplies, no US homes heated by gas are expected to go cold this winter. In some areas, however, no new residential customers are being accepted. And some utilities are telling home owners to turn off decorative outdoor gas lights.

Prices already are going up and that trend is expected to continue, even without deregulation of gas prices. Deregulation would mean even higher prices, but its backers and opponents differ on just how much.

President Ford's proposed excise tax of 37 cents per thousand cubic feet, another part of his energy package, would cost the average homeowner $70 a year, some industry officials say.

Grim Prospects

The outlook for future supplies of natural gas in this country is dismal, according to the FPC Bureau of Natural Gas. In a year-end report, it said production has passed its peak and can be expected to keep going down. From the study:

"Past efforts to effect a turnaround in the national supply posture have been largely ineffective, and we view the likelihood of success in the future with pessimism."

The Project Independence blueprint, prepared by the Federal Energy Administration [FEA], is only slightly more optimistic. It says supply will keep declining until 1980, no matter what course is followed.

But with deregulation and stepped-up production, including gas from Alaska, the FEA says current levels of production can be regained by 1985. Growth above present supply, however, is not envisioned.

IV. ALTERNATIVES TO FOSSIL FUELS

EDITOR'S INTRODUCTION

If 95 percent of present US energy consumption is of fossil fuels, what substitute sources can be found as the fossil fuels run out or become too costly? The articles in this section attempt to answer the question. The first, from *Changing Times,* surveys the most likely alternatives to petroleum. The alternatives range from familiar sources—hydroelectric dams—to exotic techniques—plasma physics and nuclear fusion.

"Atomic Power: A Bright Promise Fading?" is a report on nuclear power, which, the writer asserts, has so far not lived up to early hopes. Nuclear plants have gone into operation much more slowly than was anticipated twenty years ago and many existing plants have experienced serious operational problems. However, many experts still believe that nuclear power has promise, and in "An Energy Manifesto" thirty-two scientists issue a statement affirming their confidence.

The next three articles discuss energy sources derived more directly, in one way or another, from the sun. They are hydroelectric power, solar power, and wind power. Of the three, hydroelectric power has already nearly reached its capacity, since there are few free-flowing rivers not yet used for power generation.

Solar power, still in its infancy, seems to offer real potential, particularly in parts of the country that have sunshine nearly all year round. What is needed is the technology that will make wide-scale use economically feasible.

Wind power, like solar power, offers a virtually pollution-free source of energy. Several companies are experimenting with a very old-fashioned idea: windmills.

The final energy source discussed in this section is

geothermal energy, the kind represented by Old Faithful, the geyser in Yellowstone National Park. In several parts of the country, heat from within the earth can be transmitted to generating plants, a technique Iceland has been using for decades. John Henahan discusses ways in which this power from the earth could be harnessed in the United States.

WHERE ELSE CAN WE GET ENERGY? [1]

Ponder these facts: Nearly every American home is equipped with heat, electric lights, a stove, a refrigerator and a water heater, energy users all. Dishwashers show up in about one third of all households, and clothes dryers, used in fewer than one fifth of our homes in 1960, are now found in half. To top it off, most refrigerators on the market are big frost-free models requiring about two-thirds more energy than smaller manual-defrost ones, and most of our TVs nowadays are color models, which consume about 40 percent more power than black-and-white sets.

New buildings everywhere require higher and higher levels of lighting. Many have walls of glass that waste heat in winter and increase cooling loads in summer. Moreover, the windows are sealed so that powered ventilation is required around the clock.

Buses and trains, our most efficient transportation, have been carrying only 3 percent of intercity passengers, while 85 percent of us go in cars that have gone up in weight and down in fuel economy.

Since 1965 our voracious appetite for energy has been increasing at the rate of 4.5 percent a year. But the production of power in this country has been virtually on dead center since 1970, according to a recent report by the Ford Foundation's Energy Policy Project. Thus in 1973, when the

[1] From article in *Changing Times*. 28:25-8. Jl. '74. Reprinted by permission from *Changing Times,* the Kiplinger Magazine, July 1974 issue. Copyright 1974 by the Kiplinger Washington Editors, Inc. 1729 H Street, N.W. Washington, D.C. 20006.

Arabs abruptly reduced our oil imports, our cars, machines, lights and furnaces were using a staggering 75,600 trillion British Thermal Units (BTUs) of energy a year, while we were producing only 62,000 trillion. In just over two decades America has gone from being an energy exporter to a nation dependent on imports for 15 percent of its energy needs, including 35 percent of its oil.

So What Do We Do?

The nation's energy budget got out of balance because of a variety of domestic and international policies (or lack of them). First it became more profitable for oil companies to find oil overseas, and this they did. Then the heaviest users of electricity and gas were granted favored rates while ads shamelessly enticed us to use more energy-consuming machines and the federal government built interstate highways that encouraged high-speed, long-distance travel. Meanwhile, a lot of energy that was counted on to take care of rising needs never materialized, as rising concern over the environment delayed pipeline and nuclear plant construction and prompted shifts from coal to scarcer but less-polluting oil and gas as fuel for electrical power plants.

Well, now that we know the problem, how do we go about reversing the trend, assuring ourselves of an adequate supply of energy without wrecking the environment? There is no shortage of ideas, to be sure. Hundreds of congressional bills on energy attest to this. President Nixon has announced a goal of wiping out the need for imported energy by 1980, christening it Project Independence. Though few experts really expect this goal of zero imports to be met, it is likely that portions of the Project Independence strategy will be considered, reshaped and eventually carried out, resulting perhaps in better energy conservation methods, stepped-up domestic production of oil and natural gas, increased use of coal and expansion of nuclear energy.

Even further ahead, we will need to employ our ingenuity and wealth to harness so-called renewable energy sources, power contained in the furnace beneath the earth's crust, in the wind and sun.

Administration energy planners already have drawn up a bill of particulars that would cost $10 billion over the next five years. To centralize the management of energy research, now spread through dozens of agencies, it proposes an Energy Research and Development Administration (ERDA), not unlike the space agency that put men on the moon. The money may not be approved exactly as requested and the agency may not be created precisely as proposed, but the point is that we know what the problem is and are beginning to heed the cry of alarm at last.

Conserve We Must

With 6 percent of the world's population, we use one third of its energy, about 360 million BTUs per person annually. This could go on if we were willing and able to pay the price, of course, particularly in harm to the environment. Or we can adopt a "conservation ethic" designed to slow the increase in our demand to the rate of about 2 percent a year, at first using such obvious energy savers as slower driving speeds and lower heating levels.

Improvements in the techniques of generating, storing and transmitting electricity are also imperative. Power plants are around 30 percent to 40 percent efficient. This could be improved to 50 percent to 60 percent with magnetohydrodynamics (MHD), a process that changes coal or other fossil fuels to gas, seeds the gas with millions of tiny particles to make it a better conductor, then pushes it through a stationary magnetic field to produce electricity.

The main hope, however, lies in developing new energy sources, ones that can add to those of today that are in short supply and, eventually, replace those that may become exhausted sooner than we expect. Ultimately, all our energy sources must be renewable and pose minimal risk to the environment.

What follows here is a rundown on the prospects as of today for each of the most favored alternative fuels of tomorrow.

Nuclear Fission

The most likely alternative source right now is the nuclear reactor, which draws heat for steam generators from the energy of splitting uranium atoms. Although forty-four nuclear plants are now operating and providing 26 million kilowatts, this is just 6 percent of the nation's electrical needs and much less than proponents hoped for a decade ago. Technical problems and legitimate concern over reactor safety, handling of toxic radioactive waste and thermal pollution from water it takes to cool reactors have slowed development to a point where it takes eight to ten years to construct a nuclear plant, almost twice as long as in Europe or Japan.

Nevertheless, the nation's utilities are counting on the atom becoming a major source of energy by the century's end.

Before that, hopes rest on development of the breeder reactor, which uses plutonium and produces more fuel than it burns. That is why it is regarded by the Atomic Energy Commission as the logical hedge against the day near the turn of the century when the supply of low-cost uranium fuel may run out. There are problems aplenty, especially in handling the breeder's plutonium fuel, but if all goes according to plan and a $700 million plant near Oak Ridge, Tennessee, becomes operational, a breeder may be producing power for home heat and lights in the 1980s.

Fusion

Beyond the breeder, scientists are banking on safe, clean energy from controlled nuclear reaction similar to the sun's process of joining atomic nuclei. Fusion depends for fuel on hydrogen, which could be supplied from the oceans without polluting for thousands of years. The problems of producing power from fusion at millions of degrees Fahrenheit are monumental, but the Administration deems the concept worthy of a requested down payment of $1.5 billion over five years for research.

Coal

Fifty years ago in most large eastern US cities everybody knew about the gashouse—a sooty structure where coal was converted into gas. This process went into oblivion when cheap oil and natural gas took over, but it may be ready for a comeback. With the price of oil and gas going up and the supply down, coal gasification is gaining new friends.

The reason, of course, is that coal is the nation's most abundant fossil fuel, with identified deposits totaling 1.6 trillion tons, enough to last hundreds of years. Moreover, gasification offers a way of obtaining clean power from coal without burning it in solid form, which pollutes the air with sulfur compounds and fly ash.

About a dozen coal gasification plants around the country are testing the economic practicality of the process. Most of the plants are complex gasifiers in which pulverized coal under intense heat reacts in a semiliquid state. Streams of air, or pure oxygen, and steam are passed through the gasifier, boosting the volume of methane obtained from the coal, and then pollutants are removed.

The gas can be raised to pipeline quality by converting leftover carbon monoxide to methane with the use of hydrogen.

Until now, coal gas has been low in heating power. Much of the current research is aimed at solving this problem. With a technique called the Hygas process some experts predict commercial production by the early 1980s, taking into account that it takes five years and half a billion dollars to open a gasification plant and mining complex.

Another plan is to extract gas directly from coal while it is still underground, an idea explored by the Soviets in the 1930s. Success would reduce both the human hazards of mining and the disfigurement of the landscape caused by strip operations. Experiments so far have produced gas of low quality, but a promising technique is under study in which coal is ignited from the surface after explosives break up solids at the bottom of the seam. Then gas is collected from the bottom of the hole.

It is also possible to get oil from coal by liquefying it, a process used by the Germans in World War II. Commercial production of oil from coal is more remote than gasification. The logistics are formidable; it would take 35,000 tons of coal a day to get up to 100,000 barrels of synthetic oil. But because existing pipelines could carry it, the process has supporters.

Coal mining has never been nearly as safe and efficient as it could be. Nor is strip mining kind to the land. Plenty of things stand in the way of a comeback of coal—reluctant investors, wary miners, and a critical lack of equipment and coal cars—but the $3 billion sought by energy strategists to upgrade the role of coal would be a start.

Oil

The reserves of shale in Colorado, Wyoming and Utah alone could provide 3 million barrels of oil a day for 600 years. But when the Department of Interior in 1968 tried to lease three federally owned tracts for development, no serious bidders showed up. That was before the Arabs put an embargo on their oil to the United States. Now shale looks better, and last January Interior had no trouble leasing six tracts with a potential of 1.8 billion barrels, which would be profitable, industry says, as long as it can be sold for $6 a barrel or better.

Commercial production may begin in 1978, and by 1985, Interior predicts, we may get 1 million barrels a day, about 10 percent of the increase in total energy demand anticipated between now and then.

The big worry is what shale mining will do to the land. Producing one barrel of oil from even oil-rich shale takes about 1.5 tons of shale and leaves over a ton of waste. A plant producing 50,000 barrels a day would have a ton of waste to be disposed of every second. The Environmental Impact Assessment Project sponsored by the Ford Foundation has already warned against dumping debris and salty waste into the Colorado River, but developers, with the support of the Interior Department, assert they can handle the waste problem.

Big as oil shale reserves are, they may be equaled or even overshadowed by oil reserves of tar sands located in Canada's province of Alberta, where experts estimate there is a potential of at least 340 billion barrels and maybe as much as 800 billion.

In 1967 a Canadian subsidiary of Sun Oil started a plant there and has boosted production to about 50,000 barrels a day. Following suit is another major firm, Syncrude Canada Ltd., comprised of several US companies, which has started construction on operations capable of 125,000 barrels a day.

Sun and Wind

Energy from the sun is powerful, renewable and "free." In just fifteen minutes solar power on the sunny side of the earth equals the energy consumed by the whole world in a year. Why not use it?

There is no big problem now in collecting the sun's energy with large panels and storing it in rocks, water or special salts for use in heating and cooling homes and commercial buildings. But each system now must be custom-designed, making the process expensive. Mass-produced solar heating units could be available in five years. But the speed with which solar heating actually comes in depends on its cost in relation to rising fossil-fuel prices and whether experiments like the one at the University of Delaware involving a four-room demonstration house with roof solar cells result in competitively priced power.

Large solar panels have to be erected to face south and

angled in a way that leaves little room for architectural variation. An alternative is the central solar farm where huge collectors capture the sun's heat and use it to make steam capable of driving electrical generators. Such farms could produce power for a city of 60,000 on a square mile and still allow either conventional farming or cattle grazing.

By the year 2000, the Joint Atomic Energy Committee predicts, we will get about 6 percent of our energy from the sun; a solar energy panel of the National Aeronautics and Space Administration [NASA] and the National Science Foundation [NSF], which next year is almost quadrupling its request for solar energy research to $50 million, estimates that by 2020 the sun could provide 35 percent of our heating and cooling needs and 20 percent of our electricity.

Interest in harnessing the wind, which is really a by-product of the sun's influence on weather, has blown hot and cold over the years. Like the sun the wind can't always be counted on to be on the job. Experiments in Europe and Russia years ago did not pan out, and this country's most famous wind-powered test, using propellers weighing sixteen tons atop a Vermont mountain, was eventually abandoned. But development of lighter materials and advanced aerodynamic designs have led to proposals for floating thousands of windmills in the Atlantic and placing them atop existing transmission towers in the West. The big problem is developing an economic, reliable way of storing wind-generated electricity; a five-year program by NSF and NASA is designed to do this.

Geothermal

While natural gas and oil were cheap, underground steam and hot water never made much of a splash except in San Francisco, which gets about 400 megawatts, close to half its electrical power requirements, from generators powered by steam from geysers north of the city.

What little research there has been into geothermal steam and water will probably be increased fourfold next year. Nearly 60 million acres of federal land, much of it in the West, may have geothermal potential. Where formations of hot underground rocks exist, engineers hope to devise a way to pump water over the rocks and recapture the released heat for use as energy. Depending on who is predicting, by the year 2000 geothermal power will produce more electricity than all of the present sources or as little as 5 percent of our needs at that time.

Waste

It has been estimated that if America's 2.5 billion tons of yearly waste were burned in power plants, it could produce half of the electrical energy now being used and we would get rid of the trash at the same time. So far, perhaps twenty cities and a number of industrial companies have followed European examples in converting solid waste to usable energy, but there are drawbacks that may restrict energy from this source to only 1 percent or 2 percent of our needs. The supply of refuse fluctuates from season to season and its heat output is low. To make shredded waste burn more efficiently, it's usually mixed with coal or oil, so fossil fuels are still being consumed.

One way around this difficulty is to use bacterial decomposition of organic waste to create methane gas, which can then be burned to provide heat or generate electricity the same way as natural gas. Engineers in Los Angeles recently launched a test project to power a 300-horsepower engine with methane collected from wells drilled in a refuse dump. The engine drives an electrical generator hooked up to the city power lines and, if the project works, it could supply power to several hundred homes.

Obviously, there is no single solution for our energy problem. Conserving what we have and finding new ways to use known fuels come first. Then the alternative sources can take their places in providing abundant energy for all.

ATOMIC POWER: A BRIGHT PROMISE FADING? [2]

Reprinted from *U.S. News & World Report.*

The advent of nuclear power as a major energy source for the United States continues to be marked by fits and starts.

Many atomic generating plants have run smoothly for years. But just as many, experience shows, are plagued by poor design, sloppy workmanship and inadequate quality control. The result is one costly breakdown after another.

Some nuclear plants are "down" for repairs about as much of the time as they are actually generating power. The costs involved run into billions of dollars a year which, in the final analysis, have to come out of consumers' pockets.

Official reports from the Atomic Energy Commission [AEC] and industry sources document some recent troubles.

For example, the AEC reports that during March [1974] there were "25 forced outages" in the 31 reactors then producing commercial power. This resulted in a loss of 3,350 hours of generating time and more than 1 billion kilowatt-hours of electricity.

And, in late May, a new AEC report revealed that every reactor in the country had at least one "abnormal occurrence" . . . [in 1973].

These were unplanned power cuts with causes ranging from radioactive steam leaks, to storm damage and operator errors. One power plant, Browns Ferry Unit 1 near Decatur, Alabama, reported 65 of these incidents to the AEC in 12 months.

Record in Crisis

In January [1974]—during the height of the energy crisis caused by the Arab oil embargo—22 of the nation's nuclear plants were reported by *Nucleonics Week,* an in-

[2] From article in *U.S. News & World Report.* 76:43-4. Je. 10, '74.

dustry trade publication, to have been out of commission for part or all of the month. Sixteen of these were down for repairs or maintenance.

In spite of the many operating difficulties, the Atomic Energy Commission and the nuclear industry still feel that the atom will live up to the bright future its supporters long have proclaimed for it. All that is needed is time, they say, to work out the wrinkles of this vast and complex technology. . . .

Yet, the record thus far is disappointing. One of the strongest incentives for a utility to turn to nuclear power plants is the promised high "availability"—the amount of time a unit is producing. However, says L. Manning Muntzing, director of regulation for the AEC:

"A disturbing finding has been that the availability of nuclear power plants has fallen short of what was expected and has in fact not exceeded the availability of comparably sized fossil-fueled plants."

During 1973, the availability of the nation's nuclear facilities actually declined slightly. Says Mr. Muntzing: "Numerous forced outages, due primarily to equipment malfunctions, have been responsible for the disappointing performance we have had."

Some of the incidents that led to shutdowns also posed safety questions, as small doses of radioactivity escaped the protective "containment" area. No one has yet been injured by the radioactivity, but critics claim the industry has been fortunate thus far.

The kinds of troubles that force nuclear units out of service are shown by two case histories as related . . . by the companies' officials.

Palisades

On a Lake Michigan beach six miles from South Haven, Michigan, sits the Palisades nuclear plant—a 700-megawatt system that went into commercial operation in

December of 1971. In two and one half years, the plant has run full-tilt for only five months.

In January of 1973, engineers for Consumers Power—the utility owner and operator—discovered leaks of radioactive steam from one of the facility's two generators. The plant was shut down for seven weeks to plug the holes.

Then last August [1973], more leaks were found, this time from the second generator. The plant has not been in operation since. When engineers took apart the generator, they discovered that many of the tubes carrying water into the generator had deteriorated dangerously. A massive repair job was begun.

Just as the utility was getting ready to put Palisades back into operation in early May of this year [1974], new steam leaks were detected in the generator that caused trouble back in January of 1973. Start-up has been delayed again, perhaps to midsummer.

The shutdown of the Palisades power plant is costing Consumers Power $3 million per month.

Pilgrim-1

Boston Edison Company's nuclear plant near Plymouth, Massachusetts—called Pilgrim-1—is a 660-megawatt system that first went into commercial operation in December of 1972.

For just over a year it ran without problems. Then strange vibrations were detected inside the reactor. The AEC ordered it shut down on December 28, 1973. It has not produced electricity since.

While the unit was down for adjustment to eliminate vibrations, officials decided to carry out a partial refueling of the reactor and run some routine tests.

During one ultrasonic examination of the reactor's inner works, engineers discovered a half-inch "discontinuity" in a six-inch-thick weld on a pipe that carries radioactive water back into the reactor after it has been run through a generator. The AEC has ordered further tests on the weld.

The shutdown is costing Boston Edison $305,000 per day, or more than $9 million each month.

Inadequate quality control is cited as the cause of problems at other plants. A spokesman for the Atomic Energy Commission said that in 1973, utilities in the United States reported 861 "abnormal occurrences" in their nuclear facilities and 43 percent of these incidents had "potential safety significance."

Eighteen incidents were considered "significant" hazards and twelve of those involved release of radioactivity above permissible limits beyond the plant site boundaries, although still within AEC safety limits. Eleven releases of radioactivity in the form of gaseous iodine-131 were related to steam leaks at the Quad Cities nuclear plant near Cordova, Illinois. Another release was at the Palisades facility near South Haven, Michigan.

Quality control also has been a problem at the Vermont Yankee nuclear plant near Vernon, Vermont. It was shut down seventeen times in its first nineteen months of commercial operation.

A Chain Reaction

In 1973, the plant was ordered out of action by the AEC for the same internal vibration problems that plagued Pilgrim-1 at Plymouth, Massachusetts. While workers were locating the difficulty, two control rods accidentally were pulled from the reactor core at the same time, setting off a chain reaction. The heavy steel protective lid was off the reactor at the time.

Control rods are long cylinders of a substance that absorbs neutrons. They are made of boron or hafnium and prevent the controlled-fission reaction in an atomic plant from turning into a nuclear holocaust.

Vermont Yankee Corporation was fined $15,000 for its violation of safety regulations. This was the second time since the AEC began imposing fines in 1971 that a utility has been penalized this way.

According to Mr. Muntzing of the AEC: "Many of these occurrences would have been prevented had strong quality-assurance programs been in effect."

The AEC is working with the nuclear industry to step up quality control at the forty-five atomic plants now licensed to operate commercially in the United States.

Summer Trouble?

There is concern, however, that if these improvements in quality control do not show quick results, the nation could be in trouble this summer [1974] as consumers start using air conditioners and create new peak demands for electricity. . . .

Delays in bringing atomic units into service, as well as huge cost increases accrued in getting them built, pose still other problems.

The ten nuclear plants that had been licensed to operate by 1971 took less than six years to move from the planning stage into service, on the average. Today this lead time has stretched out to between nine and ten years. The delays are caused by prolonged licensing procedures, construction difficulties, technical problems, and intervention by groups concerned about safety and environment.

For example, the Toledo Edison Company's Davis-Besse nuclear plant was expected to start commercial service in December of 1974 on the shores of Lake Erie. Company officials now say a start-up date in early 1976 is more likely and costs have increased $60 million to a total of $415 million.

"Shakedown" Needed

The industry feels that before any final judgments are made about the capabilities of nuclear power plants, utilities should be given a chance to work some of the "bugs" out of the systems. As spelled out by the Atomic Industrial Forum, an industry association:

"Valid comparisons between the performance of nuclear power stations and fossil-fired plants are difficult . . . Most fossil plants are much smaller than most nuclear plants, and they have been operating for considerably longer than the one to five years' experience of most nuclear plants."

The Forum points out that like any other piece of machinery, atomic plants need a "shakedown" period before reaching their most efficient levels of production. Older facilities, the Forum reports, generally show improving operating figures.

Mr. Muntzing of the AEC notes, however, that 1973 figures for older plants are not anywhere near the 80 percent availability estimates many thought would be achieved by these matured facilities. He reports: "The average 1973 availability factor for plants in this older age group was 67 percent, less than the average for all plants."

The atom now accounts for about 1 percent of total US energy demand. If it is to become a major energy source that will carry the United States into the next century, say officials of the Atomic Energy Commission, it must prove itself a reliable and safe way to go.

"The enormous potential of energy from nuclear fission has been established beyond doubt," says Mr. Muntzing, but, he adds:

"Nuclear power facilities, to be safe, must be designed, constructed and operated with a discipline that has not normally been required in American industry. . . . People at every level must be made to understand that quality is everybody's business."

AN ENERGY MANIFESTO [3]

We, as scientists and citizens of the United States, believe that the Republic is in the most serious situation since World War II. Today's energy crisis is not a matter of just a few years but of decades. It is the new and predominant fact of life in industrialized societies.

The high price of oil which we must now import in order to keep Americans at their jobs threatens our economic structure—indeed, that of the Western world. Energy is the lifeblood of all modern societies and they are currently held hostage by a price structure that they are powerless to influence.

In the next three to five years conservation is essentially the only energy option. We can and we must use energy and existing energy sources more intelligently. But there must also be long-range realistic plans and we deplore the fact that they are developing so slowly. We also deplore the fact that the public is given unrealistic assurances that there are easy solutions. There are many interesting proposals for alternative energy sources which deserve vigorous research effort, but none of them is likely to contribute significantly to our energy supply in this century.

Conservation, while urgently necessary and highly desirable, also has its price. One man's conservation may be another man's loss of job. Conservation, the first time around, can trim off fat, but the second time will cut deeply.

When we search for domestic energy sources to substitute for imported oil, we must look at the whole picture. If we look at each possible energy source separately, we can easily find fault with each of them, and rule out each one. Clearly, this would mean the end of our civilization as we know it.

[3] Text from "No Alternative to Nuclear Power; 32 Scientists Speak Out," by H. A. Bethe and others. *Bulletin of the Atomic Scientists.* 31:4-5. Mr. '75.

Our domestic oil reserves are running down and the deficit can only partially be replaced by the new sources in Alaska; we must, in addition, permit offshore exploration. Natural gas is in a similar critical condition; in the last seven years new discoveries have run far below our level of gas consumption. Only with strong measures could we hope to reverse this trend.

We shall have to make much greater use of solid fuels. Here coal and uranium are the most important options. This represents a profound change in the character of the American fuel economy. The nation has truly great reserves of these solid fuels in the earth. Our economically recoverable coal reserves are estimated to be 250 billion tons and exceed the energy of the world's total oil reserves. Our known uranium ores potentially equal the energy of 6,000 billion tons of coal; lower-grade ore promises even more abundance.

The US choice is not coal *or* uranium; we need both. Coal is irreplaceable as the basis of new synthetic fuels to replace oil and natural gas.

However, we see the primary use of solid fuels, especially of uranium, as a source of electricity. Uranium power, the culmination of basic discoveries in physics, is an engineered reality generating electricity today. Nuclear power has its critics, but we believe they lack perspective as to the feasibility of nonnuclear power sources and the gravity of the fuel crisis.

All energy release involves risks and nuclear power is certainly no exception. The safety of civilian nuclear power has been under public surveillance without parallel in the history of technology. As in any new technology there is a learning period. Contrary to the scare publicity given to some mistakes that have occurred, no appreciable amount of radioactive material has escaped from any commercial US power reactor. We have confidence that technical ingenuity and care in operation can continue to improve the safety in all phases of the nuclear power program, including the difficult areas of transportation and nuclear waste disposal. The separation of the Atomic Energy Commission into the Energy Research and Development Administration and the Nuclear Regulatory Commission provides added reassurance for realistic management of potential risks and benefits. On any scale the benefits of a clean, inexpensive, and inexhaustible domestic fuel far outweigh the possible risks.

We can see no reasonable alternative to an increased use of nuclear power to satisfy our energy needs.

Many of us have worked for a long time on energy problems and therefore we feel the responsibility to speak out. The energy famine that threatens will require many sacrifices on the part of the American people, but these will be reduced if we marshal the huge scientific and technical resources of our country to improve the use of known energy sources.

HYDROELECTRIC ENERGY [4]

Hydroelectric energy is electricity produced from the energy of falling water. This is actually stored solar energy: the water being lifted from the sea and carried to high elevations in the course of the hydrological cycle, which is driven by the sun. Hydroelectric energy has several striking advantages over other current ways of producing electricity. No fuel is required, since the energy comes from the sun. There are no combustion products, no other wastes from the generation process, and there is almost no localized thermal pollution—only heat loss from friction in the turbines and generators, and electrical losses in transmission and distribution. Even the heat that eventually results from using the electricity is not an added burden on the environment because an equivalent amount of heat

[4] From Chapter 3 of *Energy*, by John Holdren and Philip Herrera. Sierra. 1971. p 52-6. Copyright © 1971 by Sierra Club. Used with permission. This chapter was written by John Holdren, a physicist at the Lawrence Livermore Laboratory.

would have been produced by friction if the water had been allowed to fall undisturbed and unharnessed. Finally, hydroelectric energy is cheap—as long as it does not have to be transmitted too far to reach its users.

The liabilities to be weighed against these assets are that usable hydroelectric sites are in limited supply, that there are ecological and esthetic costs associated with exploiting these sites, and that the eventual filling of reservoirs with silt destroys their usefulness. The first problem does not mean there is a limited amount of hydroelectric energy (the supply will last as long as the sun and the oceans do) but that there are only so many sites at which this continuously renewed energy source can be economically tapped.

The capacity of hydroelectric installations in this country at the end of 1969 was 53 million kwe, and these facilities produced 17 percent of the electricity consumed here. Thus hydroelectric energy accounted for only about 4 percent of the total US energy use. Although the production of electricity at hydroelectric plants has been increasing steadily, its fraction of the electricity budget has been falling because other sources have been growing more rapidly. The Federal Power Commission believes, based on stream flow records and surveys of feasible dam sites, that about 125 million kwe of potential capacity remains to be developed, mostly in the Pacific Northwest and in Alaska. Unless present trends change drastically, the fraction of electricity produced by hydroelectric plants will continue to drop even if the additional potential capacity is used.

Harnessing hydroelectric energy is usually just a matter of building a dam with turbines and generators installed at the base. The energy that would have been dissipated as the water descended over the submerged steam bed is then stored as potential energy in the form of water held up behind the dam. When electricity is needed, tunnels through the dam are opened so that stored energy—the high pressure of a column of water tens or hundreds of feet high—drives water through them with great force, spinning turbines which turn generators much like those in steam plants. Some hydroelectric installations have capacities equal to the largest fossil fuel and nuclear plants: there are five plants of more than 700,000 kwe on the Columbia River, and one of 2 million kwe on the Niagara.

In practice, hydroelectric energy has been much less expensive than other types. The price of electricity delivered to consumers in the Northwest, where hydroelectric units comprise more than 90 percent of all installed electric capacity, has averaged less than half the cost in other parts of the country. On the other hand, it is difficult to say what hydroelectric power would cost if it reflected the full price of the dam projects. Because such projects are often judged to have other benefits besides power production, especially flood control and water supply, the recipients of these benefits foot some of the bill. Additionally, a great deal of federal money is usually involved.

The esthetic objection to hydroelectric projects is that they turn some of the nation's most beautiful gorges and river valleys into large, dull lakes. For those who enjoy power boating, water skiing, and warm-water fishing, this may be considered a benefit. Probably the most powerful conservation argument on this point is that there are already a great many lakes where such activities can be pursued, but very few remaining gorges and wild rivers. In some instances, hydroelectric projects cover prime farmland, which eventually proves to be a poor trade. For those whose homes and livelihoods have been submerged, the personal cost of hydroelectric power is high.

Hydroelectric dams and their reservoirs may have other adverse effects, as well. The spawning grounds of migratory fishes of commercial and recreational importance, such as salmon, are often destroyed, or the fish are prevented from reaching them. The proposed Rampart Dam in Alaska would have destroyed one of the great remaining wildlife habitats on the North American continent. Seepage from reservoirs may raise the water table and bring subsurface salts and minerals with it, impairing the fertility of the soil. Reservoirs lose water by evaporation in proportion to their surface area (and depending also on local climatic conditions), increasing the concentration of dissolved minerals in the water remaining. And, in some instances, the filling of large reservoirs has triggered earthquakes because the weight of the water accumulating shifted the balance of stresses on the earth's crust.

Finally, hydroelectric reservoirs are subject to eventual filling with the silt that is carried in varying quantities by every river. Rivers such as the Colorado, which carries a particularly heavy burden of silt, may succeed in filling large reservoirs after one or two centuries. In other rivers, the process may take longer but can be accelerated by erosion from logging or upstream development. Whatever the case, the storage capacity of the reservoir is eventually destroyed. When that happens, one can still tap the energy of the waterfall that then occupies the dam site, but the available energy is then completely subject to the fluctuations in river flow.

Another process utilizing the energy of falling water is called *pumped storage*. This involves two reservoirs, a high one and a low one, connected by pipes containing dual-purpose pump turbines. During periods of low demand, such as late at night, electricity pumps water from the lower reservoir to the higher one. Later, when demand is high, the water is allowed to fall back to the lower reservoir, spinning the turbines and generating electricity. Because it takes about 30 percent more power to pump the water uphill than is recovered when it comes back down, pumped storage power makes sense only as a means of meeting peak demand. The utilities can afford it because the pumping is done when their "base-load" plants would otherwise be idle. The extra cost of running them is only the cost of the fuel—perhaps two tenths of a cent per kwhe. When the electricity is recovered later to meet actual needs in the power grid, it can be sold for three times as much.

Most environmental disruption caused by a pumped storage facility results from the upper reservoir (the lower one is usually an already existing lake or stream). The disruption usually is less than that at most hydroelectric sites because the upper storage reservoir can be relatively small. On the other hand, the visual problem may be aggravated because the water level fluctuates wildly according to how much energy is being stored. This exposes large expanses of ugly mud and discourages recreational use.

ENERGY FROM THE SUN [5]

The city of Santa Clara lies fifty miles south of San Francisco in a robustly sunny valley. As in much of California, rain is concentrated in the winter months, leaving nearly three hundred days a year of clear skies. Until now no one paid much attention to the economic value of all that sunshine. But things are changing. By July [1975] the city will have completed a new recreation building that will draw about 80 percent of its heating and cooling energy from solar collectors mounted on the roof. After that the city itself will plunge into the solar energy business. "What we see is a city-owned solar utility," says City Manager Donald Von Raesfeld. "The city will finance and install solar heating and cooling systems in new buildings. Consumers will pay a monthly fee to cover amortization and maintenance of the solar units. This will be done on a nonprofit basis, with the capital raised through municipal bonds."

Santa Clara isn't alone in its effort to convert sunshine into useful energy. A recent survey listed sixty-eight US buildings, either completed or near completion, that are

[5] From "Who'll Control Sun Power? The Solar Derby," by Peter Barnes, West Coast editor. *New Republic*. 172:17-19. F. 1, '75. Reprinted by permission of *The New Republic*, © 1975, The New Republic, Inc.

getting some or all of their energy from the sun. Dozens of corporations are involved in solar research and some are already marketing solar-related products. The federal government is pouring millions of dollars into solar research and development projects. And while the big commitment of government and industry is still to fossil fuels and nuclear fission, energy from the sun is no longer dismissed as farfetched or far off. According to a Westinghouse study funded by the National Science Foundation, solar heating and cooling of buildings will be economically competitive in most parts of the country by 1985–90, and are already almost competitive in sunny regions like California and Florida. By the end of the century, says the NSF, the sun could provide more than one third of the energy we use to heat and cool buildings, plus 20 to 30 percent of our electricity needs. It could dramatically reduce peak demands for electricity—mainly for summer air conditioning—and conserve fossil fuels for petrochemical uses for which there are no ready substitutes. Congress is equally enthused. Last year it passed five laws dealing wholly or partly with solar energy research, spreading money somewhat chaotically among the NSF, NASA, HUD [Department of Housing and Urban Development] and a new energy research and development agency.

The attractions of solar energy are apparent. It doesn't pollute or otherwise damage the environment. It creates no dangerous waste products such as plutonium. It won't run out for a few billion years. It can't be embargoed by Arabs or anyone else. It's virtually inflation-proof once the basic set-up costs are met, and would wondrously improve our balance of payments. The technology involved, while still not perfected, is much less complex than nuclear technology. And of all energy sources the sun is the least amenable to control by cartel-like energy industry.

Why then has it taken so long to discover the sun? One reason is that the energy contained in sunshine is diffuse and fickle compared to the concentrated energy found in fossil fuels. As long as fossil fuels were plentiful and fairly easy to get at, it was considerably more profitable to collect and sell these stored forms of solar energy than to capture the sun's current energy emissions. Another reason is the massive commitment of dollars and scientists the United States made after World War II to the development of nuclear energy, a commitment that in retrospect appears to have derived at least partly from guilt over having unleashed the atom for destructive purposes. ("If sunbeams were weapons of war, we would have had solar energy centuries ago," chemist George Porter has observed.) Solar energy is finally looking attractive because fossil fuels are no longer cheap, and because the drawbacks of nuclear fission—its hazards, huge capital costs, and low gains in net energy terms (it takes enormous amounts of energy to build reactors and prepare their fuel)—are now more widely appreciated.

Despite the attractiveness of solar energy, there are barriers to its introduction and use. The oil companies are much more interested in oil, coal, natural gas, uranium and geothermal steam—resources they can control and sell at hefty markups. The electric utilities don't see much mileage in the highly decentralized uses of solar energy, though they are starting to look at such centralized uses as the generation of electricity on "solar farms" in the desert, and such semicentralized uses as community-sized heating and cooling systems. The housing industry is traditionally slow to innovate, especially when innovations involve higher initial costs. So if solar energy is to be widely used before the end of the century, it must be pushed by all levels of government. The pushes need to come in four areas: improvement of solar technology, demonstrations of its practicality in different regions of the country, reform of building codes and zoning laws, and provision of incentives such as tax breaks, low-interest loans and outright subsidies. Fortunately things are moving in most of these areas.

The cooling and heating of buildings and water consumes 25 percent of the energy used in the United States, and solar technology is already proficient in capturing energy for these purposes. Back in the 1930s and 1940s, solar water heaters were installed on thousands of rooftops in Florida; installation stopped only because natural gas became available at low cost. In Australia, Japan and Israel, solar water heaters are widely used today. The basic concept is simple: water flowing through glass-covered panels is heated by the sun and piped to a storage tank. The hot water is then used directly for bathing and dishwashing, piped through the house for radiant heating, or used to power an air conditioning system. Improvements are still needed in the efficiency of solar panels (they could absorb more of the sun's energy than they currently do) and in the application of heated water to air conditioning. The federal government is funding research on both these problems and, says Lockheed's Ken Marshall, an aerospace engineer turned solar researcher, "It's just a matter of a few years before we have efficient systems adapted to each area of the country." Cost reductions are also on the horizon as companies move toward mass production. PPG Industries is now selling solar rooftop panels at about $6 per square foot; two Israeli manufacturers are exporting panels to the United States at about the same price, and General Electric is "looking at the possibility" of getting into the solar panel business. Ultimately the price should fall to around $3 per square foot.

Photovoltaic cells made from silicon have been used for years to generate electricity in space vehicles; the trouble is that their cost is high and their efficiency is low. Present costs are about $20 per watt; it should be under 20 cents per watt to be competitive with nuclear-generated electricity. Dr. Joseph Lindmayer, a leading solar cell expert, believes cost breakthroughs will come as soon as mass production starts, just as has happened with other semiconductor products. That could lead to a proliferation of solar cells on rooftops, generating on-the-spot pollution-free electricity without transmission losses. Here, too, federally funded research is playing an important role, though solar advocates say funding could be substantially increased.

Research in solar thermal conversion—using focused solar energy to drive turbines that generate electricity on a large scale—is being conducted under NSF grants by Westinghouse, Honeywell, the Aerospace Corporation of Los Angeles and a team of scientists at the University of Arizona headed by Aden and Marjorie Meinel. The electric utilities are also privately funding research in solar thermal conversion through the Electrical Power Research Institute in Palo Alto, California. The basic concept, first suggested by the Meinels several years ago, is for giant "solar farms" in the sun-drenched desert of the southwestern United States; a variant proposed by Dr. Peter Glaser, vice president of Arthur D. Little, is a solar-energy-collecting satellite, orbiting in permanent sunlight, that would beam its energy back to a receiving station on earth. Because of the enormous capital outlays involved in both approaches, commercial solar power generation is unlikely before the 1990s.

The focus for the immediate future is on solar heating and cooling. To demonstrate its practicality and to learn more about its on-site performance, the NSF has funded several solar-heated buildings in addition to Santa Clara's. As the technology improves there'll be many more government-funded demonstrations. Under the 1974 Solar Heating and Cooling Act, HUD will install solar hardware on as many as five hundred houses scattered throughout the country. The aim is to educate architects, builders, construction workers and plumbers as well as the general public. An important part of the education is the concept of life-cycle costing—including future fuel bills in calculating the cost of different heating and cooling systems.

The biggest need is for legislation creating incentives

for production and use of solar equipment, and that too is starting to happen. Florida last year passed a law requiring all new single-family residences to be designed to incorporate future installation of solar water-heating equipment. Indiana now exempts solar equipment from property taxation, and Arizona allows accelerated depreciation of solar hardware. The California Coastal Conservation Commission has recommended that all new construction in the coastal zone contain solar-assisted heating and cooling systems as soon as the necessary equipment is on the market, and Santa Clara officials are talking about a city ordinance that would do the same thing. Incentives at the federal level—a tax break equal in magnitude to the oil depletion allowance, or low-interest loans patterned after the rural electrification program of the 1930s—would dramatically accelerate the solarization of America. In Japan many local governments provide 5.5 percent loans for installation of solar equipment, and some throw in grants of up to 30 percent of initial costs.

The idea of subsidies for solar energy is being pushed by citizens' groups in several parts of the country. One such group, the Alternative Energy Resources Organization (AERO), is made up of Montana farmers and ranchers. AERO is calling for a federal tax-free production allowance for solar and wind energy similar to the oil depletion allowance, and is pressing the Montana legislature to use revenues from coal extraction to support development of nonpolluting renewable energy sources. AERO's vision was summed up by Egan O'Connor, a solar energy expert, at the Montana alternative energy conference . . . [in 1974]: "Imagine Montana telling other states to develop solar energy too instead of ripping up Montana for coal. Imagine Montana ten years from now without strip-mining, without dead rivers and water starvation, without radioactive pollution and the threat of atomic terrorism. Imagine a Montana which would still sparkle, thanks to solar energy."

As Americans look to sunshine to fill more of their energy needs, a critical question will be: who will control the sun? The answer depends on such things as control of desert land, patents and the kinds of solar technology that predominate. Fortunately there is an opportunity to develop solar energy the way Santa Clara is proceeding—by the public, on a nonprofit basis. Most of the desert lands in the Southwest are owned by the federal government or Indian tribes. Because most solar research is government funded, the government can license nonexclusively most of the advances in technology. This could lead to a highly competitive solar hardware industry, similar to the semiconductor and building supply industries. If the most common ways of capturing sunshine are decentralized—i.e., rooftop panels and photovoltaic cells—then a significant portion of America's energy could be owned by and could go directly to its users, without passing through corporate hands.

On the other hand, there are ways the energy corporations and utilities can get hold of the sun. At least three oil companies—Mobil, Exxon and Shell—are conducting research in photovoltaic cells, presumably with an eye toward controlling the eventual market through patents. The electric utilities are starting to think about how they might lease desert land and generate electricity through centralized solar thermal plants. (If the Southwest desert is used for solar electric generation, it ought to be done by a "solar TVA.") Another approach being considered by utilities is the "renting" of solar heating and cooling equipment, much the way the telephone company "rents" telephones. Thus a utility would retain ownership of the system and charge a monthly rate that would cover amortization and maintenance plus a profit.

Utilities might also profit from the sun through alteration of existing rate structures. Because there's no known remedy for bad weather, decentralized solar energy systems must plug into centralized electric or gas systems for back-up. Utilities argue that if solar energy becomes widespread, they'll still have to maintain big load capacities for periods of inclement weather, but won't be able to defray all their capital costs through sale of energy in fair weather. Therefore, they say, they'd be entitled to a "commitment charge," based on how much back-up energy a consumer "reserves," in addition to an actual usage rate.

The way solar energy is controlled will probably be determined within the next decade. The stakes are high, since energy from sunshine will be a multibillion-dollar-a-year business by the end of the century. Donald Von Raesfeld, Santa Clara's city manager, believes "the sun is one resource the public ought to get hold of as soon as possible." At a [January 1975] conference of the National League of Cities . . . he urged cities to take the lead by getting into the solar utility business themselves. So far, however, very few cities, states or federal agencies have given much thought to how to keep the sun a public resource.

PULLING POWER OUT OF THIN AIR [6]

If you watch television regularly, you have probably seen this commercial from Atlantic Richfield, the company that discovered the Prudhoe Bay oilfield on the North Slope of Alaska: (Scene: Windmill, its sails turning slowly in the dusk, lights shining from its windows.) "Back in 1915, three thousand windmills helped light up the country of Denmark. America could generate electricity the same charming way. All we'll have to do is keep wasting our natural stores of energy. When it's [*sic*] all gone, we'll just turn on the windmills. A great idea, till the wind dies down." (Scene: The windmill's sails stop turning, the lights go out, and someone starts cursing the dark in Danish.)

Or, if you are a videophobe, perhaps you have read

[6] From article by Gary Soucie, a conservationist who writes frequently of environmental topics. *Audubon*. 76:81-8. My. '74. Reprinted from *Audubon*, the magazine of the National Audubon Society; copyright © 1974. Reprinted by permission.

this ad by the Caterpillar Tractor Company, the makers of earth-moving and coal-mining equipment, in the pages of *Business Week* and other influential magazines: (Scene: Two men along the bank of a canal. The younger one, his longish hair carefully coiffed in that artsy-faggy Madison Avenue way, and wearing a flared-lapel sportcoat over *purple* pants and shirt, points at a windmill across the canal and says, "The perfect energy. Clean, cheap, and can't foul up the environment." The other man, older and more prosperously stout, his short hair in an Iowa National Guard crop, and wearing sensible shoes and a no-nonsense windbreaker, his probity and pensiveness symbolized by the briar pipe in hand, says, "But it can't meet today's needs.") The text of the ad begins, "Too bad about windmills. They just aren't a practical solution to our energy dilemma." After a paean to coal the ad concludes, "There are no simple solutions. Only intelligent choices."

Maybe it's just environmental paranoia, but it does seem eerily coincidental that these public attacks on windmills and wind power by oil and coal-mining interests come hard on the heels of recently revived scientific interest in harnessing the kinetic energy of the earth's atmosphere. In 1970, Stewart L. Udall devoted one of his syndicated newspaper columns to windmills, a column so popular it was rerun in 1972. "Windmills are much, much more than relics," Udall wrote. "They are symbols of sanity in a world that is increasingly hooked on machines with an inordinate hunger for fuel and a prodigious capacity to pollute. Ecologically, the windmill is one of the few perfect devices. It harnesses a completely free resource to pump water or generate electricity under conditions that respect the laws and limits of nature." . . .

It Began With the Persians

Wind power isn't exactly a new idea. So far as we know, windmills were invented in Persia over a millennium ago. The earliest historical references are to a Persian millwright in A.D. 644 and to windmills in Seistan, Persia, in A.D. 915. When Genghis Khan's hordes swept through Persia, captured millwrights were taken back to China, where to this day windmills are still used to pump water for irrigation. . . .

For six and a half centuries, from the late twelfth through the early nineteenth century, windmills were widely used in Europe. Wind and water powered the dawning of the Industrial Revolution. Then along came James Watt and his patented steam engine, and suddenly you could have a mill town without strong winds or running water. For a hundred years the use of windmills declined slowly but surely. After World War I the widespread use of electricity and internal combustion engines led rapidly to the virtual demise of the windmill.

In 1890 a man named LaCour built a mill in Denmark that used the power of the wind to generate electricity; but again, water could do it better, and the coal- or wood-burning steam turbine better still. The use of windmills persisted to a degree in areas where there were no streams to be dammed or no ample supplies of fuel. Three or four decades ago, rural homesteads all across the United States used windmills to generate electricity as well as to pump water for washing and drinking, for cropland irrigation, and for watering livestock. But the federal government had a better idea, and the massive rural electrification program relegated the windmill to an anonymous piece of the "other sources" category in the statistics of electric power production.

Power from Grandpa's Knob

Still, against all these odds, wind power almost made a comeback between 1939 and 1951. Just prior to the outbreak of World War II in Europe, a Yankee entrepreneurial combine of engineering, manufacturing, and utility companies —recognizing the virtual exhaustion of hydropower sites and worried about the cost and availability of coal in New

England—got together to build a great mother of a windmill on Grandpa's Knob near Rutland, Vermont. It was the biggest windmill ever built—not one of your quaint European gingerbread houses with sails, either, but a 175-foot-diameter two-bladed propeller mounted on a steel skeleton tower and powering a turbine generator that produced 1,250 kilowatts of electricity. It was, one might say, a very substantial pinwheel.

In 1940, after it had been in operation only a short time, the Grandpa's Knob windmill ground to a halt, having burnt a bearing. It was repaired, but some four years later a second breakdown—this time, one of the propeller blades—brought the experiment to an end. Economic studies had shown that wind power was only marginally competitive with steam power plants. And between the end of the war and the early 1950s, coal prices dropped sharply and the nation found itself with an excess of electrical generating capacity. (Not all the news from Grandpa's Knob was bad. That broken propeller blade led to great advances in metallurgical engineering.)

In 1943, Vannevar Bush, the physicist who was director of the wartime Office of Scientific Research and Development, became worried over dwindling American fuel reserves and concluded that wind power might be an answer. Dr. Bush did two things that were destined to come to naught. First, he installed a wind power advocate named Percy H. Thomas as a staff consultant to the Federal Power Commission. Second, he got the War Production Board in 1945 to sponsor wind power research at New York and Stanford universities. The NYU–Stanford project wind-tunnel-tested two aerodynamic designs based on prewar wind machines and, even with the technological constraints of the day (the use of synchronous-speed shafts and blade design limitations forced by the state of the art in metal forming), got "mediocre to excellent" results.

Meanwhile, Thomas had convinced the Federal Power Commission that the Department of the Interior should build a large prototype wind power plant. But the idea died aborning in the House Committee on Interior and Insular Affairs in 1951, an innocent victim of the Korean war. That year, hard-pressed to finance both guns and butter during the first of our undeclared wars in Asia, Congress had to pass not one but three supplemental appropriations bills. . . .

Both Feasible and Impractical

Meanwhile, back on the Continent, things were not faring much better. Denmark, Germany, France, Italy, and Spain all conducted wind power experiments, producing results that were judged feasible but impractical.

Prior to World War II, Denmark had developed a network of small windmills with a total generating capacity of 100,000 kilowatts. After the war, it occurred to the Danes that most of what little power they had during the Nazi occupation came from jerry-built wind generators. So Denmark launched a wind power research program. The results were promising, but the Danes abandoned large-scale wind power generation in favor of buying electricity at discount rates from Sweden, which has a surplus of hydropower potential but needs paying customers to justify its development. The old Danish mills are still working, having run trouble-free for nearly half a century with virtually no maintenance of any kind. . . .

By 1950, a great many planners in Great Britain had decided that wind power, in combination with tidal power, was the answer to Britain's postempire fuel starvation predicament. But after eight years of fruitful research on the Welsh and Cornish coastlines, where the wind patterns and sea cliffs fit the conventional wind power theories, the plan was dropped in favor of the premature promise of unlimited cheap power from nuclear fission.

Wind power schemes on both sides of the Atlantic suffered from common weaknesses. First, their economics were not measurably better than those of fossil-fueled power plants. Second, none of them provided for adequate storage of the power that could be generated during windy periods for use during the summer doldrums. So it was assumed they would have to be paralleled 100 percent by some other generating capacity. And third, wind power seemed to have been made obsolescent by nuclear power, which promised cheaper power and more high-technology fun. . . .

They Come from the Sun

Winds are produced by solar energy. The differential heating by the sun of the earth's surface causes a lateral heat flow that keeps the particles of the global atmosphere in motion. Other winds are caused by the evapotranspiration cycle, heat being stored in the atmosphere, until it rains and cools things off, allowing the heat buildup to start all over again.

It is hard to say just how much kinetic energy in the atmosphere is available to earthbound windmills. The World Meteorological Organization, using an extremely conservative approach, in 1954 calculated that 20 billion kilowatts of power are available for harvest at very specially suited sites around the world, mostly bald mountain knobs and sea cliffs in regions of very high average wind velocity. Other estimates of available wind energy range as high as 80 trillion kilowatts in the Northern Hemisphere alone. That's a lot of power. According to the Federal Power Commission, in 1970 the total world electrical generating capacity was 1 billion kilowatts, of which the United States alone generated and consumed nearly one third.

Wind energy is not equally distributed around the globe, but it is remarkably consistent within large areas. Generally speaking, the major wind systems intensify from equator to pole, with much modification of these patterns by the relation of land and water masses, topographic features, and other factors. Besides the global wind systems—the equatorial doldrums, the trade winds, the westerlies, and the polar easterlies—there are at least fifty-two local wind systems extensive enough to have been given names. Some of the more well known are the sirocco, monsoon, mistral, bora, foehn, helm, tramontane, simoon, and the southerly buster of New South Wales. While the experts may quarrel over numbers, there is no disputing the enormity of the kinetic energy resource in the global atmosphere.

All that energy just blowing around, going to waste, vexes William E. Heronemus, professor of civil engineering at the University of Massachusetts and undisputed pretender to the throne of King Aeolus, the warden of the winds in Greek and Roman mythology. Heronemus' involvement in Aeolian Edisonry began in 1970 when he and thirty-two colleagues began investigating nonpolluting power systems. Discouraged that the best near-term bet seemed to be floating atomic reactors, he decided that maybe the time was ripe for reviving the wind power work started on Grandpa's Knob. . . .

Windmills at Sea

As envisioned by Heronemus, the New England Offshore Wind Power System (OWPS) would be a gigantic network of windmills on floating platforms or concrete-pile Texas towers out on the Georges Bank and on Nantucket and New York shoals. The windmills would be used to generate electricity to run offshore electrolyzer stations for producing hydrogen gas from sea water. The hydrogen would then be piped ashore and converted, by fuel cells or direct combustion, into electricity. Heronemus sees the hydrogen link as the key to OWPS. Hydrogen can be stored in undersea tanks, can be piped without energy loss, and can be converted into virtually pollution-free electrical energy.

The entire Offshore Wind Power System would include eighty-three wind units, compressor and deep-sea storage systems, an offshore collection subsystem, and shoreside terminals, distribution subsystems, and fuel cell substations. Each wind unit would comprise 164 wind stations (each with three two-bladed windmills powering generators of

600- to 2,000-kilowatt capacity) arranged in concentric rings around an electrolyzer station, which would sport its own three-wheel array.

The whole system could, according to Heronemus, generate 159.2 billion kilowatt-hours of electricity per year—which is exactly the projected 1976–90 increase in electrical demand for New England. At a capital investment cost of $22.3 billion, OWPS is a bit steep for your average local electric utility company. But projected fossil-nuclear costs are in the same range, and Heronemus expects that expected improvements in fuel cell efficiencies will make OWPS economically—as well as ecologically—superior to the dig-drill-and-burn power generation methods now in use.

Heronemus sees only minimal environmental impact from the Offshore Wind Power System. Inevitably, there will be some disturbance of the sea bed during construction, but none during operation. At present there is no market for the huge quantities of oxygen that would be generated as a byproduct of electrolyzing water for hydrogen, so it would probably be bubbled into the surrounding waters, which should enrich the fisheries resource. . . .

A Nuclear Road to Ruin

The concern for New England's economic future is one of several reasons Heronemus cites for his paternal interest in wind power. In most polls of industries that don't want to move to New England, the high cost of energy ranks high as a reason, right behind the sky-high tax rates and cost of labor. New England is utterly dependent on the sheiks of Araby and the oil barons of the Gulf Coast and Texas for her energy, and Heronemus is convinced that the nuclear route is a one-way road to ruin. . . .

New England's winds are not the only ones that have attracted Heronemus' attention. He has been working up charts, tables, graphs, and formulas for other regions of the United States that show wind power promise. According to Heronemus, the whole northern third of the country has winds strong enough and consistent enough to produce electricity. And along the East Coast and in the Great Plains, the wind-swept regions extend as far south as the Carolinas and Texas. Heronemus is particularly keen about harvesting the winds on Long Island ("Long Island could get every bit of her energy from wind power"), New Jersey ("you could go along the Garden State Parkway with hanging arrays of small-diameter wind mills that could be as attractive as suspension bridges or the Gateway Arch in St. Louis"), down the axis of the Great Lakes with one exception ("Lake Michigan's winds aren't much"), and the northern Plains and [Great] Lake States ("Wisconsin, Minnesota, Michigan, Montana, and the Dakotas could harvest a valuable second crop from windmills along fence lines and above the forests").

Nor is Heronemus alone in looking to the winds for electrical energy. At Oregon State University, an interdepartmental team under Dr. E. Wendell Hewson, professor and department chairman of atmospheric sciences, has been studying the power potential of the winds along Oregon's shoreline and up into the Columbia River Valley. Dr. Hewson and his colleagues are less visionary than Heronemus and regard wind power as supplementary to, rather than replacing, conventional power plants, possibly as early as 1980–85.

"If it were possible to install additional generators in dams to use water pumped back into reservoirs by wind power," Dr. Hewson explains, "you could use the existing system without building more dams and further endangering the environment. Raising the level of a reservoir by just a few feet could make quite a difference for hydrogeneration."

Someday-Maybe Schemes

Moreover, H. Guyford Stever, director of the National Science Foundation, recently told a congressional committee that it would be possible to put windmills atop skyscrapers to supply part of the electricity for the cities below. But wind power investigations in the back rooms of academia can't produce kilowatt one. The energy companies see wind power as one of those someday-maybe schemes advanced by academics, environmentalists, counterculturalists, and utopians of every ilk. The conventional ledger-book wisdom is that wind power is *interesting,* but impractical, inefficient, and uneconomic.

To this kind of thinking Heronemus snorts, "In the United States today, 'impractical' usually means that something isn't fancy enough to appeal to our ideas of high technology and flamboyant life style. 'Inefficient' usually takes on a more scientific and quantifiable meaning that in the last analysis means, 'I know how to achieve a higher carnot or coulomb efficiency with a different process.' The engineer looks for the most efficient process because it is usually the most economic. But if we are seeking pollution-free energy, that guideline is not necessarily the best one to follow. 'Uneconomic' means that something cannot compete in the marketplace. Solar energy schemes cannot compete in the marketplace so long as ground water is a 'free good' heat sink for heat engines, so long as our air is a 'free good' refuse dump for combustion or fission airborne emissions, and so long as our earth is a 'free good' repository for high-level nuclear wastes." . . .

Pinwheel Esthetics

Putting windmills offshore of New England or Oregon or down the middle of the Great Lakes also dodges the esthetic issue, but what about rigging hundreds or thousands of 800-foot windmills or pinwheel arrays on the flatlands of south Jersey, the Great Plains, or the Lake states, or on the flanks of the White or Green Mountains in New England, the Koolau Range on Oahu, the coastal foothills of Alaska, or the other ridges and ranges coveted by wind power enthusiasts?

Says Wisconsin native Heronemus: "I consider myself a good conservationist and have been all my life," citing his National Wildlife Federation membership as proof. "I'm a farm boy and I love the earth and nature. But if you admit that we must have a certain level of energy to keep our civilization going—and I myself don't see how we could reduce our current average usage by more than 20 percent—then we must have additional power plants or, hopefully, new types of power plants that will remove the pollution caused by those now in operation. If one accepts that, and then realizes that the winds in certain wind-swept regions could supplant all of the existing fossil-fueled or uranium-burning plants, then shouldn't one be far more willing to look at windmills once in a while?"

(In the same interview, however, Heronemus was explaining how the University of Massachusetts campus could be wind powered: "The cheap and easy way to do it would be to put several arrays along the top of the Holyoke Range. But even I am not sure I'd like to see anything up there. I look out at the Holyoke Range from the breakfast table.")

European windmills have never seemed an esthetic blight, but Heronemus' engineering drawings of efficient towers and poles of steel and aluminum atop which are perched aerodynamically severe nacelles housing the generators to which are attached huge two-bladed propellers don't seem destined for inclusion in snapshot albums. It would seem that a bit of architectural artistry could make them esthetically passable, at least in the silo and grain elevator austerity of the plains. . . .

Wind Power When?

If wind power is such a good thing, you may be wondering, why aren't we reaping the winds already? First, there is the matter of marginal economics. But as fossil-fuel prices continue to rise out of sight, as the true cost of generating nuclear power becomes so evident that even the "nuke" nuts stop talking about cutting the cost of electricity in half, wind power will look more attractive to the keepers of the corporate ledger. Next, some manufacturing moguls have got to see the potential windfall in making oversize pinwheels. As Heronemus succinctly puts it, "Nobody makes the hardware yet, but that's more plus than minus. If New England would get with it, she could free herself from the fuel merchants, reassert her former economic independence, get all the energy she needs without pollution, and develop a new industry that could provide the manufacturing jobs she so badly needs." And once, but probably not until, the conversion efficiencies of fuel cells have been improved by maybe 40 or 50 percent, all the hydrogen-linked solar power schemes will start gathering momentum.

And there is a subtler problem. As a society, we still don't seem to realize just how much trouble we are in, and least of all the electric utility companies. The energy crisis that broke last fall [1973]—the same one we "Malthusian prophets of doom" have been warning against for a decade or so—seems to have even more firmly convinced the utilities, the White House, and the suckered news media that nuclear power is The Answer. The electric utility industry is in the throes of a flaming romance with isotopes and reactors. . . .

If wind power were a new and untried idea, if it weren't an old-hat idea and a Persian old-hat idea at that, the utilities might be falling over themselves to try it out. Instead, they are knocking themselves out, and maybe the rest of us along with them, trying to make light-water reactors work and can hardly wait to get their hands on the liquid-metal fast-breeder reactors. Nothing alarms a technofreak so much as to suggest that he look backward or toward the inscrutable East.

Someone—I wish I could remember who—once said that at the brink of an abyss, the only progressive step is a backward one. Just think of the sociotechnical revolution that might overtake us if we started looking backward—not only to windmills, but also solar heaters, steam and electric cars, passenger trains, houses built with more regard for climate than for architectural fashion, products designed for use rather than for throwing away.

Looking backward while reflecting upon wind power, one is reminded of a line Shakespeare wrote in *Henry VI:* "Ill blows the wind that profits nobody." Or as my grandmother used to put it, "It's an ill wind that blows no good."

GEOTHERMAL ENERGY[7]

Today everyone agrees that geothermal energy is an abundant and essentially unlimited power source. The only problem: devising systems that can turn it into electricity, and do it efficiently.

All early geothermal fields—such as those at the geysers in California; in Laradello, Italy; and in New Zealand—are so-called dry-steam reservoirs. But those are rare: They exist only where a supply of underground water comes in contact with cracked hot rock and is turned into steam. Where this happens, it's simple to get geothermal electricity; just drill a hole and pipe the natural steam to a nearby turbine.

But there are two other potential sources of geothermal power, and these are plentiful: hot-water and hot dry-rock deposits. Enormous amounts of energy are stored in such reservoirs—but it's not so easy to get it out. Some hot-water deposits—such as the vast reservoir that underlies California's Imperial Valley—are not pure water but violently corrosive brines that can clog up drill holes or eat out machinery in days. Getting energy out of hot dry-rock presents considerable difficulties of its own.

All the experts agree: Before we can get really large amounts of energy from the earth, we'll have to solve a lot of tough technological problems involved with hot-water and hot dry-rock fields. . . . Research teams at work in these areas . . . [are] coming up with a lot of ingenious solutions,

[7] From article by John F. Henahan, a science writer. *Popular Science.* 205:96-9+. N. '74. Reprinted with permission from *Popular Science* © 1974 Times Mirror Magazines, Inc.

and . . . they're excited about the prospects of geothermal energy. As they talk, you can sense the potential payoff that excites them.

Fantastic Amounts

Dr. Don White, of the United States Geological Survey in Menlo Park, California, estimates that the heat in the top ten miles of the earth's crust totals 3×10^{26} calories. That is about 2,000 times the amount of heat that would be produced if we burned the world's entire supply of coal. Much of that heat is so spread out or so deep below the earth's surface that it defies commercial exploitation. It has been estimated, however, that if only one tenth of the geothermal energy in the top two miles of the earth's crust could be extracted by today's techniques, and converted to electricity, it could provide 58,000 mw annually for at least the next fifty years. Dr. Robert Rex, president of Republic Geothermal in Whittier, California, and one of the foremost pioneers in the field, estimates that there is enough energy sealed beneath the Imperial Valley alone to meet the electrical needs of the Southwest for at least two hundred years.

A report issued by a committee headed by former Secretary of the Interior Walter Hickel estimates that it should be possible to develop as much as 132,000 mw of geothermal generating capacity in the United States by 1985. On the other hand, the more conservative National Petroleum Council foresees that geothermal power capacity in the same period will reach no more than 3,500 mw. Within these extremes, geothermal energy could provide from 0.5 to 20 percent of the nation's electrical power within the next eleven years.

How They'll Do It

Basically, there are three ways to use geothermal hot water to produce electricity. In Cerro Prieto, Mexico (in the southernmost extension of the Imperial Valley), the Mexican government has been operating a 75-mw geothermal power plant that uses the flashed-steam process. . . . In this system, pressure reduction as the hot water rises in the well causes some of it to flash into steam. The steam runs a turbine that generates electricity.

Another way to take advantage of geothermal hot-water potential is through the "binary vapor" technique now being developed by Magma Power Company in Los Angeles. In this case, the fluids are not allowed to flash into steam; instead a pump in the well keeps them under high pressure, and the hot fluids are fed into a heat exchanger where they vaporize a low-molecular-weight fluid such as isobutane or Freon.

In the Magmamax process, the vapor drives a low-pressure turbine, then is condensed and recycled through the heat exchanger. Within a few months, the Southern California Edison Company plans to use the Magmamax process in a small 10-mw power plant that will get its energy from the hot-water geothermal deposit beneath Mammoth, in northern California.

Unfortunately, neither system uses much of the heat present in the geothermal water. The flashed-steam process uses only the part that flashes into steam. The heat in the remainder is unused. The binary vapor technique does not use the geothermal fluid directly, a necessarily less efficient system. The geothermal hot-water deposits in the Imperial Valley cannot be used directly in a power generation system because they are really brines—with salt concentrations ranging from about 2 percent to as high as 25 percent in the Salton Sea area. Corrosion and caking caused by the more concentrated brines make it almost impossible to use them in a conventional power-generating system.

Nevertheless, a group of never-say-die researchers at the University of California's Lawrence Livermore Laboratory find a lot of room for maneuvering in that "almost." They believe that a new type of "total-flow impulse turbine" they are developing may be able to produce electrical power from even the most concentrated brines in the valley. . . .

The total-flow system—the third and possibly most efficient way to use geothermal hot-water deposits . . . is so named because it will run on both the liquid and steam phase of the fluid mixture that comes out of the geothermal wellhead. In principle, it is related to the impulse turbines that have been used in hydroelectric power plants for more than a century. Water from a dam or some other source is expanded through a spray nozzle; the force of the droplets acts against the scoop-like vanes of the turbine wheel, which then turns to operate a generator. . . .

By 1978, if all goes well, a small one-megawatt power plant using a total-flow impulse turbine should be feeding electricity into the grid of a power company in southern California . . . By 1981, it should be possible to scale up to a more impressive 200-mw size.

Higgins is confident that a geothermal power plant could be built and operated for about the same or even less money than traditional power plants.

"It all depends on how difficult it is to solve the materials problem," . . . [Higgins said]. "But even if the turbine material is as expensive as titanium, we should be able to produce electricity for about 7 to 13 mills per kilowatt-hour. Power plants now fueled with coal or oil come in at around 12 mills/kwh, while nuclear plants will probably cost about 12 to 15 mills/kwh by 1980, just about the time when geothermal energy should be coming into its own." . . .

While the Imperial Valley's hot brines are undoubtedly a geothermal bonanza, there is an even greater supply of untapped heat stored in concentrations of hot dry-rock formations—usually granite—relatively close to the earth's surface. It has been estimated that 95,000 square miles of a thirteen-state area in the American West are underlaid at a depth of about 3.5 miles with hot dry-rock averaging temperatures of about 550° Fahrenheit (290° Centigrade). During the last five years, Morton Smith, Don Brown, and Bob Potter, all of the University of California's Los Alamos [Science] Laboratory (LASL), have had a strong yen to do something with all that heat. They devised what sounded like an outrageous idea at the time: to pump water down into the hotrock area and create useable steam in the same way that nature does. . . .

The LASL researchers planned to use a new rock-melting technique they had just developed to drill a well deep enough to reach dry-rock areas where the temperature is about 450° Fahrenheit. Since then, they decided that conventional oil well drilling techniques would be just as economical and equally efficient.

Once the well is dug, water is pumped down the shaft to create a hydraulic fracture in the rock, a pancake-shaped network of interconnected cracks and fissures. The next step is to pump more water into the fracture zone, wait for it to heat up, then withdraw the steam from a second well drilled into the upper edge of the fracture zone.

On the Brink of Success

In April 1973, at least part of the plan became reality. On the edge of a volcano near Los Alamos, the LASL team created a circle-shaped fracture about 140 feet in diameter at a depth of 2,500 feet. The rock temperature was "a very satisfying 110.4° Centigrade," according to Morton Smith. Everything went as predicted. Since then, with that mild success under their belts, he and his colleagues have been digging other wells, shooting for deeper depths, larger fracture zones, higher temperatures, and granite formations tight enough to hold the hot water. . . .

If the LASL hot dry-rock technique works in New Mexico, Smith sees no reason why it could not also be applied to the Conway granites beneath New Hampshire or even the rock underlying New York's Manhattan Island. However, those formations will probably not be hot enough to produce steam for electrical power, he says. Instead, they could be used to produce water hot enough to be used in space heating or air conditioning systems. Reykjavik, Iceland, which sits on a geothermal hot-water deposit, uses 275° Fahrenheit water for heat and hot water.

Prospects and Plans

Meanwhile, the search for new sources of geothermal energy or new ways of using the old sources goes on:

☐ In Marysville, Montana, a team headed by researchers from Battelle Memorial Institute in Richland, Washington, is sinking a well into what could be a multibillion-dollar geothermal hot spot. The thermal reservoir, covering a ten-square-mile area, was apparently created by the intrusion of hot lava from the earth's mantle into its crust tens of thousands of years ago. Temperatures at the bottom of the well are 7,000° to 8,500° Fahrenheit. This first well is primarily a scouting probe designed to find out just how large and hot the deposit is. Exactly how the subterranean energy supply will be converted into electrical energy depends on whether it turns out to be an impermeable hot dry-rock area as expected, or a hot-water deposit. Drillers should determine this soon. Any prospects for commercial development are at least two years off, says project manager Bill McSpadden. The three-year project is sponsored by a $2.5 million grant from the National Science Foundation.

☐ The Bureau of Reclamation plans to use the Imperial Valley's geothermal brines as a source of fresh water to replenish the Colorado River's dwindling and increasingly salty lower reaches. A 30,000-gallon-per-day pilot desalination plant is now operating near Holtville, California. It operates like the conventional desalting plants, except that it isn't necessary to heat the brine, which rushes from the wellhead at 300° Fahrenheit.

☐ Researchers at the University of California, Riverside, have mapped the geothermal deposits of the Imperial Valley by measuring tiny electrical currents in the ground—a method that's faster and cheaper than drilling test wells.

☐ The Atomic Energy Commission is investing $8 million to construct an experimental 10-mw power plant over a geothermal hot-water reservoir near Battle Mountain, Nevada.

☐ The Geysers dry-steam field north of San Francisco continues to prove its worth. A new power plant, to be finished in 1977, should increase the capacity of the installation from its present 396 mw to 502 mw.

☐ The federal government has finally begun to lease hundreds of thousands of acres of federal land to private geothermal developers.

There is no doubt that a geothermal bonanza lies somewhere beneath our feet, and the trend of geothermal research and development is definitely on the upswing. Once new energy sources are discovered, they will certainly be put to good use.

V. THE ENERGY OPTIONS

EDITOR'S INTRODUCTION

Intelligent choices have to be made in the mid-1970s if the lights are not to flicker in the 1980s. The choices pose such questions as these: How much should the United States emphasize energy saving and how much should it emphasize developing new fuels? How shall energy research funds be allocated? Should the United States fund a crash program to develop nuclear energy or should it fund more research on solar energy? Should the government mine the oil shale it owns? Should federal air pollution standards be lowered to permit the burning of more coal? Should taxes on gasoline be increased steeply to discourage driving?

The articles in this compilation make it evident that there are few clear-cut answers to these questions. Well-intentioned people disagree sharply on the specific remedies to be applied. This final section explores some of the decision making processes involved in attempting to find solutions to the problems. First, there is a look at what the federal government is doing, in a summary of a government report on Project Independence. The report, prepared by the Federal Energy Administration, recommends curbing the growth rate of energy consumption from about 4.3 percent a year to 2 percent a year. The report assumes that by about 1985 nuclear energy will be supplying 30 percent of the nation's electricity.

The next article, by Marilyn Wellemeyer of *Fortune,* presents the possible decisions President Ford can make on energy, emphasizing the fact that hard decisions are long overdue. The selection that follows describes a new federal superagency, the United States Energy Research and Development Administration. ERDA's mandate is a large one:

direct and coordinate all the government's activities in energy research and development.

An article by Russell Peterson, chairman of the President's Council on Environmental Quality, outlines the so-called Half and Half Plan developed by the CEQ. The plan calls for equal measures to conserve energy and to increase energy production in order to bring supply into line with future demand. Peterson argues that the United States must become self-sufficient in energy—that it must stop buying energy abroad.

In a somewhat similar vein, the next article summarizes a massive study commissioned by the Ford Foundation on energy choices. The study urges a major effort to conserve energy, including the channeling of funds away from energy development and into research on increasing the efficiency of existing sources.

The concluding article introduces one more complication into the discussions of future sources of energy. Wilson Clark argues that with many of the proposed new energy sources, so much energy is consumed in the process of creation that in the end there is actually only a very little net gain. We should *use* less, he contends—an argument that has been sounded again and again.

PROJECT INDEPENDENCE [1]

Following are highlights of the Project Independence report . . .

Conservation

A determined effort, including mandatory federal standards for cars, appliances and buildings; industrial planning; more public transportation, and redesign of electricity rates could cut energy growth to 2 percent a year by 1985, less than half the recent growth rate. Demand-management

[1] From "Highlights of the [Project Independence] Report." New York *Times*. p 68. N. 13, '74. © by The New York Times Company. Reprinted by permission.

measures, such as prohibiting natural gas and oil heating in new buildings, could shift demand to electricity and diminish oil imports and vulnerability to another embargo.

Oil

Domestic United States production of crude oil, which peaked in 1970, will continue to decline for several years, despite 1973–74 price increases, because of the time needed to explore and develop new fields. At $11 a barrel, 1985 production could range from 15 million to 20 million barrels a day, depending on government policies. The naval petroleum reserve in Alaska could yield 2 million barrels a day. If the price drops to $7 a barrel, 1985 production could range from 12 million to 17 million barrels a day.

Natural Gas

Annual consumption is two or three times greater than annual discoveries. Reserves have been falling since 1967. If federally regulated prices for gas moving in interstate commerce remain at present levels (43 cents per 1,000 cubic feet for new wells), "the outlook for increased gas supplies is not promising," particularly for deliveries to nonproducing states. Higher prices could lead to large additions to supplies onshore and from the offshore Gulf, Atlantic and California coasts. Alaska's naval petroleum reserve and continental shelf could also make significant contributions, as will the North Slope.

Coal

From supplying 90 percent of the nation's energy at the turn of the century, coal has fallen to 17 percent. Only electric utility consumption has been rising. It now accounts for two thirds of coal use. At the 1973 production rate, 599 million tons, the United States has 800 years of coal reserves, half of it east of the Mississippi and high in sulfur. Almost any foreseeable 1985 level of production is possible at or below present coal prices.

Electricity

Nuclear power has become increasingly important, but is well behind schedule for various reasons. The Federal Energy Administration forecasts nuclear plants will provide 30 percent of all electricity by 1985, a forecast lower than many. Overall future electricity requirements are highly uncertain. Redesign of rates could diminish peak loads and need for new generating capacity.

Shale Oil

Production could reach 250,000 barrels a day by 1985 at $11 a barrel. Shale is concentrated in Colorado, Utah and Wyoming. Environmental costs of development would be heavy, as would be requirements for water in a water-scarce area. Development might also require federal price support.

Solar Energy

Despite high capital costs and storage problems, technological advances and rising fossil-fuel costs "will enable solar energy to become a commercial energy source," but it will not be significant compared with other sources until after 1985. Cost reductions will depend more on volume than on technological advances, and consumer acceptance could be slow. Solar energy could significantly reduce pollution, but it may pose land-use and esthetic problems.

Geothermal Energy

The United States has vast sources of heat in the earth, but development is just beginning and has been retarded by delays in award of federal leases on western lands under 1970 legislation. Contributions to energy supply by 1985 will probably be slight but might be substantial by the turn of the century. Tapping the earth's heat does relatively little environmental damage, confined to the generating site, but does require the stringing of power transmission lines.

PRESIDENT FORD'S HARD CHOICES ON ENERGY [2]

After four years of dither and drift, the United States government at last has begun coming seriously to grips with the energy crisis. When President Ford sets forth his energy policy . . . [in January 1975] in a message to Congress, he can scarcely avoid making numerous difficult decisions that have been inordinately delayed, at immense cost to both the United States and other countries.

Despite four energy czars, a five-month Arab oil embargo, and a massive study of the problem by the Federal Energy Administration, a coherent energy policy so far remains as elusive as when former President Nixon first ignored the warnings of his chief economist, Paul McCracken, who counseled in November 1970, that plans must be laid for dealing with inevitable future shortfalls in energy supplies. Until very recently, Ford has shown no more sense of urgency than his predecessor about coping with difficulties that can only increase with procrastination. Indeed, it was not until December 7 [1974]—nearly four months after he assumed office—that Ford submitted to his first full-scale briefing on energy problems, a two-hour session at the White House with a dozen top advisers.

"Complex as the Devil"

The floundering pace of policy making is understandable, up to a point. At that belated briefing, according to Press Secretary Ron Nessen, President Ford called the situation "complex as the devil." It is certainly that. More important, rarely have the stakes been so high, or the policy alternatives so politically unpalatable.

By abruptly quadrupling the price of oil that most nations cannot do without, at least for some years, the Arab

[2] From article by Marilyn Wellemeyer, staff writer. *Fortune*. 91:75-7+. Ja. '75. Reprinted from the January 1975 issue of *Fortune* magazine by special permission; © 1974 Time Inc.

oil cartel has accelerated global inflation, increased the threat of starvation in some underdeveloped countries, and jeopardized the solvency of the industrial West. Barring a military response to the OPEC [Organization of Petroleum Exporting Countries] brand of economic warfare, the only fundamental remedy is to beat down the high cost of foreign crude by peaceful means. And the United States, the world's largest oil and energy consumer, must take the lead by adopting stern measures to conserve oil and develop more domestic energy supplies. Unless we adopt "a policy of austerity," warns Arthur Burns, the chairman of the Federal Reserve Board, there may be "a permanent decline of our nation's economic and political power."

As is usual when hard economic choices must be made, the pain in any effective program that President Ford proposes will come almost immediately and the payoff much later. Decreasing energy consumption could slow an already weak economy and increase unemployment in some industries (tourism, for example). Reducing oil imports will require either gasoline rationing, allocation, or new taxes on petroleum.

Measures to boost domestic energy production will seem, to many people, almost as unpleasant. If the output of domestic oil and natural gas is to be increased quickly and efficiently, price controls will have to be ended—jolting consumers with a rise in gas and oil costs. Environmentalists, an increasingly powerful political bloc, can be counted on to oppose the needed expansion of offshore drilling and of strip mining out west, where most of the low-sulfur coal lies. To make use of our abundant supplies of coal elsewhere, air quality standards will have to be lowered. Some coal men insist, for example, that restrictions against high-sulfur coal must be relaxed for twenty *years* in order to give operators an adequate incentive to open new mines.

Deterred by the Downside Risk

One of the difficulties is that many new domestic energy sources, including offshore and Alaskan oil, yield comparatively high-priced fuel. Shale oil or synthetic crude from coal might cost $10 or $12 a barrel. Years of effort and billions of dollars of investment will be necessary to bring either to the marketplace. To meet all our domestic energy needs over the next decade, says Treasury Secretary William Simon, may require a total investment of $850 billion.

Private companies would be foolish to commit vast sums of capital to costly long-term energy ventures at a time when there is widespread hope of a sharp break in oil prices within four or five years. In this regard, a fundamental but often overlooked fact weighs in the calculations of energy companies. Once oil has begun flowing from a Middle Eastern well, production costs are only about 10 cents a barrel. Persian Gulf producers can, if they choose, cut their prices drastically and still make a lot of money. As a consequence, those expensive forms of domestic energy won't be developed unless the United States Government provides some kind of protective umbrella to shield producers from the potentially ruinous competition of cut-rate Persian Gulf oil.

What We Can, and Can't, Afford

There are, in short, no easy, comfortable solutions to our energy predicament. So the nation must sort out the true priorities. To begin with, we can certainly afford more costly energy than we have had, though not painlessly. . . . freeing all domestic crude from price controls would directly add only about 0.7 percent to the GNP [gross national product] deflator, roughly a third of what the OPEC cartel has added already. And we can afford some degree of reliance on foreign oil over the long run, provided the sources are diversified and that we have insurance against another embargo in the form of a stockpile, or increased standby productive facilities.

But what the United States—and even more so Europe and Japan—cannot afford is continuing to go into debt at the present rate to purchase foreign oil. . . . [In 1974] foreign fuel cost the United States $26.5 billion, more than triple the $8.3 billion we paid for about the same amount of fuel in 1973. That bill turned our otherwise healthy trade surplus into a deficit of more than $5 billion. But we import only 38 percent of our petroleum, and 16 percent of our total energy.

The plight of Europe and Japan is far worse, for they must import nearly all the oil and most of the energy that they consume. The strain is particularly evident in Japan, where consumer prices have leaped 26 percent in a year while the pace of the economy has dropped below its 1973 level.

At the present rate of price gouging, the OPEC countries are taking in some $110 billion a year, and spending less than half for imports. At these prices, according to estimates by the Paris-based Organization for Economic Cooperation and Development (OECD), the oil cartel will accumulate assets of $425 billion by 1980. That is enough to buy up all the world's gold, or nearly seven eighths of the common stocks traded on the New York Stock Exchange, at recent prices. And even if the real price of oil were allowed to erode by a third by the end of 1980, the cumulative OPEC surplus might fall only to about $375 billion.

"A Crisis of Capitalism"

The awesome international consequences of this transfer of wealth seem to be perceived only dimly by most Americans. Assistant Secretary of State Thomas O. Enders has portrayed the prospect in somber detail. "Unless they act," he told a congressional committee not long ago, "the industrial democracies face an inexorably rising danger of financial collapse, or depression, or both over the next decade . . . It is impossible that Europe, Japan, and America could undergo a decade of threatening financial collapse and low or no economic growth without the most shattering social and political upheavals . . . It is no accident that the Soviet Union and China, securely self-sufficient in energy, with a sustained growth rate, have begun to analyze and exploit a great new crisis of capitalism."

Diplomacy and moral suasion having failed, there is not much hope of driving down the OPEC price unless consuming countries take joint action to curb their rising appetite for oil. The basic strategy, under the plan put forth by Secretary of State Henry Kissinger in mid-November [1974], calls for consuming countries to impose a collective zero growth rate on their oil imports for the next ten years, cutting oil imports from the present one third to one fifth of their total energy needs. That, combined with the market pressure from new oil finds in such countries as Mexico, Peru, China, and Malaysia (which may add ten million barrels a day to world production within a few years), might break the cartel price.

But consuming countries seem unlikely to reach that goal unless the United States curtails its own oil imports sharply. Among the major oil importers, only the United States and (thanks to North Sea oil) Britain have any prospect of attaining self-sufficiency in energy within a decade. Accordingly, the Kissinger plan calls for the United States to reduce oil imports from an anticipated seven million barrels a day this year to one million by 1985.

That is a stunningly ambitious target, and hitting it will require that President Ford make not just one but a number of those unpopular decisions so long delayed. To be sure, the President in October [1974] called on the nation to reduce oil imports by one million barrels a day this year. But two months later, he was still insisting, the nearly unanimous opinion of his energy advisers to the contrary, that voluntary conservation was "making headway" toward that goal. Again and again, he rebuffed suggestions for a higher federal tax on gasoline.

By contrast, European leaders have been low-keyed in

their pronouncements but quicker to take unpleasant, practical steps. While Ford shrank from decisive action, French President Giscard d'Estaing announced that his country will use import controls to keep its 1975 oil imports at . . . [1974's] $11-billion level. Britain increased the tax on gasoline to 67 cents, raising the retail price to $1.20 per gallon. New taxes boosted the price in Italy by 23 cents, to $1.72.

Lost in a Regulatory Maze

There are disturbing signs that, in its gropings for energy self-sufficiency, Washington will continue to neglect the most effective weapon available: hitching market forces to what we want to happen. . . . [In December 1974] the conservation measures receiving the most attention included an import quota and the petroleum allocation program that had meant gasless Sundays, alternate-day fill-ups, and endless lines at gasoline stations during . . . [1973's winter] embargo. Some industry experts who were called to Washington to offer advice to Administration officials professed to be "appalled" at the low level of enlightenment shown by the questions they were asked.

As the massive Project Independence study forecasts, energy consumption should remain quite sensitive, over time, to changes in price. By deregulating the price of oil and natural gas, the government at one stroke would curb consumption (because prices would rise), provide a powerful incentive for efforts to increase the supply, and restore market efficiencies to an industry now trapped in a maze of disruptive regulation.

Much of that maze, of course, results from decades of misguided congressional zeal for consumer protection and, more recently, the notion that petroleum profits in particular are "obscene." In controlling the price of natural gas, as it has been required by a court decision to do for the past twenty years, the Federal Power Commission [FPC] has kept the price so low as to discourage new exploration, and create a needless shortage. Now the FPC is finally letting the gas price go up—too little and too late—from a ceiling of 42 cents to 51 cents per thousand cubic feet. The price will have to rise some more before we stop squandering our cleanest and most convenient fuel under industrial boilers, and put natural gas into its rightful market relationship with other fuels.

There is just as strong a case for junking the clumsy two-tier system that has governed domestic crude oil prices since August 1973. The beguiling idea was to keep oil companies from collecting windfall profits by putting a ceiling on the price of "old oil" (in general, crude they were producing in 1972) while encouraging them to lift their domestic output by letting the price of "new oil" rise to world market levels. The price of old oil remains frozen at an average $5.25 a barrel while new oil now brings more than $10.

One top Administration official acknowledges that the system is partly responsible for the 2.2 percent decline in domestic oil production last year, but argues that any effort to decontrol old oil prices would have impelled Congress to impose a too-low price ceiling on *all* domestic crude. In that case, of course, production might decline even further, and new exploration certainly would. Under the circumstances, President Ford . . . [in December 1974] signed into law a bill extending the present controls setup through August [1975].

Sitting on "Old Oil"

The Administration's political judgment may be correct, but the two-tier system is turning into an economic and bureaucratic monstrosity. The Federal Energy Administration [FEA] is floundering with the complex job of allocating everybody a fair share of the cheaper oil. The $5.25 ceiling prompts some operators to sit on their old oil rather than produce it, in expectation of eventually getting a higher price. In some cases it may have encouraged feverish drilling, on leases adjacent to old wells, for oil that could be classified as new—drilling that has only worsened the shortage of steel tubing and other oil-producing gear.

By keeping that $5.25 price level on old oil, we sacrifice some—perhaps a large part—of the huge but unrealized potential for secondary and tertiary recovery of more oil from old wells. Primary recovery, which includes pumping, generally extracts only about a quarter of the crude in an oil field. The use of secondary recovery methods, generally involving water or steam flooding, raises the average recovery to about one third. In many fields, the technique is necessary to keep production from declining over time, but under price controls the $5.25 price applies to that portion of secondary recovery needed to restore a well's 1972 level of output. Secondary recovery adds anywhere from 25 cents to several dollars a barrel to the cost of producing oil, depending on geologic conditions. So its use normally hinges on the price of the oil. At $10 a barrel, a lot more secondary recovery becomes profitable than at $5.25.

The cost to consumers of scrapping price controls on oil would be comparatively modest, probably adding 5 cents to the recent average gasoline price of 53 cents a gallon. Under ordinary circumstances, ending the unpopular oil depletion allowance might lift the price by another 2 cents a gallon, but the impact would probably be nil so long as the world price of oil remains sky-high. As for those much maligned "excess" profits, the right remedy is not price controls but carefully designed taxes that would encourage pumping profits into new exploration and production. . . .

Congress should also override state production controls with a law compelling the unitization of oil fields. Under that arrangement, each oil reservoir is run as a unit with all producers sharing costs and profits. If compulsory unitization were combined with price decontrol to encourage more secondary and tertiary recovery, we might even see a 35 percent rise in domestic oil output within three years, from 8.9 million barrels a day to 12 million. . . .

An Ominous Deadline on Coal

Coal production must be increased from . . . [1974's] level of some 600 million tons to more than one billion tons a year by 1985, so that utilities and big industries can make a major shift to coal as their primary boiler fuel. The FEA reckons that, by then, if all power plants stop using oil, we can save 1.7 million barrels a day, or more than a fourth of our present imports. But Congress must act swiftly to clear away uncertainties about pollution controls if the goal is to be reached.

As matters stand now, nearly half of the coal supplies used by the nation's utilities may become unusable on July 1 [1975], because the fuel will fall short of sulfur-emission limits set by states under the 1970 Clean Air Act. That law should be amended, as the Ford Administration's Energy Resources Council has proposed, to postpone the deadline ten years. Congress should also give the Environmental Protection Agency [EPA] power to relax emission standards that states have enacted on their own. In many cases these are stiffer than the federal ones.

The bitter battle between the EPA and some utility companies over stack emissions creates a further obstacle to increased coal production. The utilities want to avoid retrofitting existing plants with scrubbers, which they contend are unreliable. But the EPA seems adamant in refusing to let coal users rely permanently on intermittent control systems already in operation, or tall stacks that keep ground-level concentrations of sulfurous gases low. While the row rages, utilities and industries are loath to sign long-term contracts without which coal men won't expand operations.

Even if all the uncertainty about pollution controls is resolved, the coal industry will face enormous problems in expanding. Among the mines opened since 1960, only twenty-three were producing more than two million tons annually in 1973. To reach 1.3 billion tons a year by 1985, the industry would have to open about 270 mines that large or larger—one new deep mine and one surface mine every month. And beginning at a time when the industry is shy of both middle management and straw bosses, it would have to recruit 80,000 new miners in the East and 45,000 in the West.

Steps to Stop Waste

At best, expanding energy supplies takes time—four years, for example, to bring an offshore oil well into production, and five or six for a new underground coal mine. The quickest way to reduce oil imports is to burn less of the stuff. The place to start is where we waste the most petroleum: motoring. The private auto consumes more than half the energy expended on transportation. Short of rationing, the most effective way to get motorists to cut pleasure driving and unnecessary trips is a big increase in the federal gasoline tax, now 4 cents a gallon. Secretary of the Interior Rogers Morton [who is now Secretary of Commerce] . . . estimates that a 10-cent increase would cut gasoline consumption about 3 percent the first year and 10 percent by the fifth year, while a 30-cent increase would reduce consumption by 8 percent the first year, or 530,000 barrels a day.

Last year the average US motorist drove 10,000 miles and got 13.7 miles to the gallon. His gasoline bill totaled about $400. A 20-cent tax increase would add $146 to the retail price of his gasoline, raising his total bill to about $546. And a 30-cent tax increase would add $219 a year, bringing the total bill to $619.

Most of the $27 billion collected from this 30-cent "conservation fee" should be returned as tax rebates, which should especially favor the poor, and those who must use cars or trucks at work or unavoidably drive long distances to their jobs. To wring inefficiencies out of our freight system, which gulps 20 percent of transportation energy, the Interstate Commerce Commission should revise its regulations to promote pooling of truck freight in remote areas and encourage private carriers to eliminate empty return trips.

Heating and cooling our homes consume the equivalent of 4.5 million barrels of oil per day. By keeping thermostats at 68° by day and 60° at night we could save one sixth of that by 1977. Installing storm windows and doors, caulking, and insulation would save about as much, and ought to be encouraged by letting homeowners deduct part of the cost on their tax returns.

By going still further, and adopting such measures as a national thermal standard for new homes and offices, commercial lighting standards, and a mandatory twenty-miles-per-gallon standard for autos, the FEA calculates that our oil savings could rise to 3.7 million barrels a day by 1985. Under some circumstances, that alone would get us about a third of the way toward Kissinger's ambitious goal for cutting oil imports.

Time to Take Risks

For the years beyond 1985, as oil and natural gas supplies perhaps dwindle or become too costly to be used for a number of purposes, coal will have to bear a heavier share of the energy load. The trick will be to use our 500-year supply without unduly befouling the environment. Substitute technologies such as coal gasification and liquefaction could come into play—provided that the country is willing to help shoulder the risk of the large investment that will be required. Liquefaction plants producing enough feed stock for one 200,000-barrel-a-day refinery might require more than twenty coal mines producing a million tons a year each. To encourage such costly projects some Administration officials have aired the idea of erecting a variable tariff wall around the United States to keep imported energy priced at the equivalent of about $10 or $11 a barrel of oil.

It is the wrong approach. If the OPEC cartel's price breaks, the tariff wall would saddle the United States with needlessly high costs for all its energy, to the detriment of the entire economy. It would be cheaper for the nation if the government just financed some costly projects itself, as it does, for example, public housing. But that might not be necessary. Borrowing the techniques by which the Federal Housing Administration piled up $585 million in profits over thirty-six years, the government could simply insure the money invested in the plants.

In the unlikely event that foreign oil is cheap and plentiful in, say, 1990, the government would have to share the losses on coal conversion. But as a measure for the economic defense of the nation, the risk is well worthwhile.

Nuclear power plants have fallen far short of their promise of cheap and ample power. Fifty-five nuclear plants are in operation, 58 others have construction permits, and 110 more are on order. But they are expected to generate only 30 percent of our electricity by 1985, owing to operating problems and construction delays. Again, the regulatory maze is one of the bottlenecks. Congress could help by adopting a bill, drafted last year by the Joint Atomic Energy Committee, to cut the red tape in siting and licensing.

"Our Liberty Is at Stake"

Self-sufficiency in energy is a long way off, even for the resource-rich United States. Ill-advised and uncoordinated policy making has probably retarded the development of domestic energy resources more than any other cause. In the bygone era of cheap energy, the mess aroused little concern. But now, untangling the web has become a matter of critical national importance. Time has run out for dawdling over studies, or balking about the last 1 or 2 percent of environmental purity. With the money that we spend for foreign oil in a month, we could open 40 two-million-ton underground coal mines or build a hundred offshore oil production platforms in the Gulf of Mexico.

Despite the growing interdependence of nations, events have not yet repealed the historical axiom that says no country remains powerful unless it controls its sources of food and energy. The vision of Henry Kissinger provides perhaps the greatest hope so far that the United States will rise to its severest test of will in a generation. As he says, "It is our liberty that in the end is at stake, and it is only through the concerted action of industrial democracies that it will be maintained."

ERDA—NEW ENERGY AGENCY [3]

In a town where Scrabble with agency names often becomes a governmental game for concealing problems, a new bureaucratic conglomerate has emerged with a broad, still nebulous mandate for solving the nation's energy problems.

Its name is United States Energy Research and Development Administration. But for everyone from the telephone operator to its top official it has quickly become ERDA, the name by which it will undoubtedly be known so long as it survives—a lifetime that some of its critics already are predicting will be relatively short.

ERDA . . . is either a bureaucratic nightmare or a bureaucratic dream, depending upon whether one believes a single agency can successfully coordinate and direct all the government's myriad activities in energy research and development.

But even at the outset for an infant agency, one thing is apparent: Probably never before has a single government agency been given such urgent, yet disparate, responsibilities. ERDA is responsible for developing everything from more efficient car engines to atomic bombs, from windmills that will generate electricity to atomic power plants that will create more fuel than they consume, from solar panels that will heat and cool buildings to plants that will turn coal into gas.

[3] From "Focusing on Energy—Unclearly," by John W. Finney, staff writer. New York *Times*. p F 3. Mr. 9, '75. © 1975 by The New York Times Company. Reprinted by permission.

The birth of ERDA in itself was somewhat of a surprise. For years the executive branch had been talking of creating a new energy development agency only to be blocked by the Joint Congressional Atomic Energy Committee, which was protecting its fiefdom; the Atomic Energy Commission [AEC].

But then in his waning days of power before retirement, Representative Chet Holifield, the California Democrat who once rode high as chairman of the Atomic Energy Commission, relented and allowed the legislation creating ERDA to pass. Some say he was worn down by Senator Henry M. Jackson, who had his own ambitions as chairman of the Senate Interior Committee. . . .

Lock, stock and barrel, ERDA took over all the functions, personnel and laboratories of the AEC except for the regulatory responsibilities which were given to a new five-man Nuclear Regulatory Commission.

With its 6,000 employees, seven laboratories and a $1-billion weapons program, the former AEC is far and away the largest component in ERDA and one of the questions is whether it will come to dominate the new agency. For balance, and out of principle, ERDA also took over the fossil-fuel research programs and the Bureau of Mines programs of the Interior Department; the solar, thermal and wind research of the National Science Foundation, and research on new automotive power systems and fuels from the Environmental Protection Agency.

The conglomeration was a natural outgrowth of the Energy Reorganization Act of 1974 that specified the new agency was to bring together all federal activities in energy research and development so as to provide a coordinated and effective development of all energy resources.

When it came to ERDA's mission, the legislation establishes a mission so broad and ambitious that one wonders whether it is supposed to create new energy by fission or fusion. The law sets forth goals for the new agency of effective action to develop and increase the efficiency and reliability or use of all energy sources to meet the needs of present and future generations, to increase the productivity of the national economy and strengthen its position in regard to international trade, to make the nation self-sufficient in energy, to advance the goals of restoring, protecting and enhancing environmental quality, and to assure public health and safety.

ERDA officially began this task on January 19 [1975], with Robert C. Seamans, former deputy administrator of the National Aeronautics and Space Administration [NASA], former secretary of the Air Force and most recently president of the National Academy of Engineering, as its administrator.

During his years in NASA and the Air Force, the 56-year-old Mr. Seamans acquired a reputation as a cautious, methodical administrator, more an engineer than a free-wheeling politician. As could be expected, Mr. Seamans, although obviously deeply immersed in politics, is bringing the same methodical approach to the solution of the nation's energy problems.

"We are not embarking on any crash program," explained Mr. Seamans. . . .

As long-range energy goals, Mr. Seamans does not talk about "independence" or "self-sufficiency"—terms still loosely and rhetorically cast about by a President and his Cabinet secretaries. Rather, his research goals are to develop, and conserve, energy resources so that by 1985 the nation will end its vulnerability to economic disruption by foreign suppliers and so that by the end of the century the United States will have the ability to supply a significant share of the energy needs of the Western world.

With a headquarters staff of 7,300 persons, another 90,000 in laboratories and a budget of $4.3 billion, ERDA organizationally is divided into six departments: fossil energy; nuclear energy; environmental and safety; conservation; solar, geothermal and advanced energy systems; and national security.

Each of the departments will be headed by an assistant

administrator appointed by the President and confirmed by the Senate—and therein lie some political problems for Mr. Seamans.

With the presidential appointments, there is an obvious danger that political interests will infiltrate the organization —something that seldom happened in the AEC staff below the commission level. There is also the danger that each of the departments will evolve into its own bureaucratic fiefdom, a possibility that Mr. Seamans recognizes and which he says he is "assiduously trying to avoid." . . .

Within the once powerful atomic energy community, there already are complaints that ERDA has a bias against atomic energy—a feeling compounded by the fact that for its permanent headquarters the agency has chosen a building on Capitol Hill near Union Station—about as far as it can get from the AEC building some thirty miles away in Germantown, Maryland. Some of the top AEC administrators complain they were "left twisting slowly in the wind" and have retired.

As might be expected, Mr. Seamans denies that there is any prejudice in the new agency against atomic energy. But at the same time he emphasizes that if the nation is to solve its energy problems, "We must go down several routes, rather than just concentrating on a few, such as nuclear energy."

"We are never again going to have a cheap energy situation, and we have got to use every string in our bow if we are going to maintain the life style of this country," he declared.

From Mr. Seamans' comments, it was apparent that ERDA is not going to be as aggressive in pushing atomic power as was the AEC, which ran into mounting criticism that it was both a promoter and regulator of atomic power plants.

The regulatory function has now been delegated to the Nuclear Regulatory Commission, headed by William A. Anders, a former astronaut who became an AEC commissioner.

Despite its newness, the commission is already trying to avoid the pitfall of many a government regulatory body—becoming proindustry in its outlook.

For the immediate future, Mr. Seamans sees ERDA's "big push" concentrated on developing uses of coal, "if only because that is something we have lots of." The agency's budget for fossil-fuel development will jump by 60 percent to $311 million, with emphasis placed on developing processes for liquefying and gasifying coal.

As the government's energy research arm, ERDA will be doing work in its own laboratories and contracting with industry and universities. But in contrast to most government agencies which contract for a finished product to be developed for them, ERDA is in the business of developing products and techniques that will be taken over by industry.

In this distinction, Mr. Seamans sees one of the most important, still unresolved policy issues facing his new agency.

"Everything we do is for naught unless it finds its way into commercial use," he maintained. "But the ways to carry this out are not clear."

One approach would be to enter into cost-sharing arrangements with industry, such as the construction of demonstration plants. But this could revive the old public versus private power fight that plagued the atomic power program in its early days.

It is a problem pressing in on ERDA in two specific areas. The projected cost of a pilot plant for a breeder reactor—toward which utilities have contributed $250 million—has now jumped from $700 million to $1.7 billion, and ERDA must find some way to continue financing the project. And then there is the billion-dollar question of whether government or industry, independently or jointly, should build the new plants for enriching uranium into fuel.

Around Mr. Seamans' temporary office are pictures of rockets and astronauts—mementos of the days when he was a key figure in the project to land a man on the moon—

and on a couch-side table is a small brass statue of Don Quixote.

To Mr. Seamans, his new assignment is far more challenging and difficult than the task of landing a man on the moon, if only because it is more diffuse. As for Don Quixote, he stays around because "I have always sort of liked him."

A LONG-RANGE PLAN FOR ENERGY [4]

Some people still think the shortage of energy last winter was a fluke, but in truth it was the public's first exposure to a problem which will dominate our national life until the end of the century and beyond. Major choices must be made about our energy future, and each of us—as voters, consumers, employees, homeowners—has a stake in and some influence over future energy policy.

The most important aspect of that policy is the goals we decide to work toward. At the Council on Environmental Quality (CEQ) we have developed a set of goals which we call the Half and Half Plan—a specific and manageable program of energy conservation and production, geared to energy needs *and* environmental needs. Like all plans, it involves numbers, graphs and projections. But implicit in the numbers and curves—at the heart of the plan—are the really important things: human needs, individual life styles and basic values.

To begin with, there is the question of whether our nation should seek to become self-sufficient in energy—dependent largely upon our own resources. Some believe such independence is neither desirable nor practical. I believe it is both.

The United States, with 6 percent of the world's population, consumes about a third of the world's energy. In the

[4] From "Nation's Chief Environmental Adviser Offers a Long-Range Plan for Energy," by Russell W. Peterson, chairman of the President's Council on Environmental Quality. *Smithsonian*. 5:80-5. Jl. '74. Copyright 1974 Smithsonian Institution, from *Smithsonian* magazine July 1974. Reprinted by permission.

rest of the world, hundreds of millions of people are living on less than $100 a year. Many nations are trapped by an explosion in their population which nullifies efforts to improve their level of economic well-being. Historically, population growth has slackened as per capita income has increased. But to accomplish economic growth requires raising the use of energy. For this reason, if for no others, the United States has an obligation to plan its energy future so that energy supplies outside the United States can be devoted to improving the quality of life elsewhere around the globe.

To rely primarily on our energy resources, however, does not mean that the economy must squeal to a painful halt or that our comfort and convenience must decrease. Energy use will continue to grow. But the question is: How much? What will our society's energy demands be in the year 2000?

Traditionally, future energy demands have been calculated by projecting current growth rates in consumption. From 1965 to 1970, energy use per capita grew at a rate of more than 3 percent a year. In 1972, the Department of Interior began with that and projected a total of 192 quadrillion BTUs (British thermal units) of energy in the year 2000, more than two and one half times present consumption. (A quadrillion is a million billion; in the jargon, it is often shortened to "quad.") This projection was an excellent technical effort, and was representative of a number of "high-range" forecasts of that period. But the oil embargo changed our perspective. That amount of energy production is simply out of the question without unacceptable dependence on foreign sources or unacceptable damage to our environment.

What must be understood is that the advertisements which read "a nation which runs on oil cannot afford to run short" should have read "a nation which runs on oil is certain to run short." Seventy-five percent of the US energy today comes from petroleum and natural gas. But 1970 was the peak year for domestic petroleum production; oil from Alaska and the outer continental shelf will merely slow the decline in what is available to us. And natural gas supplies are expected to reach their peak shortly. From here on out, domestic oil and gas production will be declining at the same time that our energy needs will be increasing. Hence, to more than double energy consumption while using only our domestic resources would require us to quadruple or quintuple present coal production in the United States, and to increase nuclear power from less than one quad today to as much as 60 to 80 quads by the year 2000.

Even if we *could,* we do not want to do that. The production and use of energy is the single most important cause of environmental degradation—all the way from the devastation of lands by surface mining and the hazards of deep mining to oil spills, sulfur emissions and other pollution of air and water. To consume 192 quads in the year 2000 would put terrible strains on the environment.

So for many reasons we need a lower goal for total energy consumed. The Half and Half Plan represents that—a target of 121 quadrillion BTUs instead of 192. We at CEQ believe this is a more desirable goal for the United States.

Half Growth, Half Conservation

The Half and Half Plan grew out of analyses of the growth in per capita energy consumption since 1947, not just during the last few years. Over the longer period the rate of growth in energy consumption per capita averaged 1.4 percent. That is lower than the recent 3 percent rate, but it supported, over the period, a great growth in national affluence.

Furthermore, the Half and Half Plan assumes that one half of this growth in energy consumption per capita can be achieved through energy conservation—through reallocation from wasteful uses to ones which meet important human needs. To paraphrase Benjamin Franklin, a BTU saved is a BTU earned. Each BTU saved for one purpose means one more that can be put to wise use somewhere else. In a nation that has been profligate in using energy, there is plenty of room for relatively painless conservation.

I believe the United States should make it a major national objective to make energy savings of 0.7 percent each year. By the year 2000, conserving at that rate will reduce gross energy inputs by 27 quadrillion BTUs, more than one third of our present energy consumption. By making such savings, only 0.7 percent of the overall growth rate of 1.4 percent must be provided through additional production of energy. Half conservation and half growth—hence, the Half and Half Plan.

The energy savings can be achieved by means . . . which . . . are familiar: better insulation and more energy-conscious architectural design in residential and commercial buildings; personal efforts to reduce household energy use; more efficient appliances; more recycling of materials; better land use; more energy-conscious design of industrial processes; and, most important of all, a more efficient transportation system.

At CEQ we have looked particularly closely at the transportation sector and think it would be possible to provide all needed transportation services in the year 2000 with 21.6 quads, a small increase above the 17.0 quads we use today. Some may feel this is heretical (the Department of the Interior, for example, projected 42.7 quads, nearly twice as much), but we think it could be done.

Improved mass transit would be an important factor, but the major change would come through smaller, more efficient cars. It is obvious that a car that gets twice as many miles per gallon uses half as much gas. (It also cuts air pollution.) If, over the next decade, we adopt a new ethic about what constitutes a desirable automobile, we can provide for universal auto ownership in the year 2000 and still consume little more gasoline than we do now.

Of course, many Americans have deep-seated feelings

about cars; many still believe a big car is an important sign of success. A few months ago I went to a meeting at the State Department. All sorts of dignitaries pulled up in front of the building in limousines, but when I arrived in CEQ's Pinto, we were waved away by the doorman. Clearly a member of the establishment would not be found in such a car! I had to park a few blocks away and walk to the meeting. The point is that we are going to have to change our attitudes about some things. But we've done it before. People were delighted with the black Model-T before color, fins, accessories and yearly model changes appeared.

Hope from New Technology

Even if such conservation is achieved, even if half of our 1.4 percent growth rate is accomplished by energy saving, a great deal of energy will still need to be produced. The question is: How can it be done in the face of dwindling supplies of oil and natural gas?

Many people have high hopes for new technologies such as solar energy, geothermal energy and nuclear fusion. I support an aggressive pursuit of these technologies, but it takes wishful thinking to expect that more than 3 percent of our total energy will come from such sources by the year 2000.

Nuclear fusion as a source of energy still faces many unknowns. Even with major breakthroughs in fusion research (and these are by no means certain to occur), we cannot expect any significant production from commercial fusion power plants by the year 2000.

Solar energy is now quite properly being pushed very hard. Its main application will be to heat and cool buildings. But solar energy will be used almost exclusively on new buildings—few old ones will be retrofitted—and it will be of limited use in the high-density urban dwellings which constitute a steadily larger percentage of new home construction. Once you do the arithmetic, you have to conclude that solar energy appears unlikely to contribute more than 1 percent of our energy in the year 2000.

The story is similar for geothermal energy, for it can supply no more than 2 percent of our electricity needs by the turn of the century.

Therefore, even if energy from the earth's heat, the sun's radiation and the fusion reaction are pushed along as fast and as far as possible (as they should be), they simply cannot solve our basic problem in the next twenty-five years, and we are left, inescapably, with the alternatives of coal and nuclear fission.

The United States has the world's largest supplies of coal and the most advanced nuclear technology. Both sources also have severe environmental problems associated with them . . . Yet environmentalists must realize that when they oppose a particular solution to a basic social need, they must have an alternative. At present, these are the only alternatives in sight, so it behooves us to establish and enforce the proper environmental regulations for both of these sources.

With coal, how to mine it is the first problem. Everyone has seen pictures of the ravages of strip mining, but not everyone knows that strip-mined lands—with the exception of relatively arid lands—can be reclaimed at reasonable cost. A study CEQ did a year ago showed that full restoration for Appalachian strip mines was achievable for about 3 to 9 percent of the value of the extracted coal. As the value of coal rises, that percentage becomes even smaller. We must insist on full restoration.

Deep mining is environmentally less damaging but is extremely hazardous to the miners. Many coal companies have a horrible record in protecting the health and safety of the men underground, but others have good records. We need to develop new mining technologies and insist on the best safety practices.

However it is mined, we must burn coal in a way that minimizes the threat to our health from emissions of sulfur oxides. Fortunately, about half of our coal has low-sulfur content; this should be allocated to densely populated areas. Furthermore, we must refine technologies for removing sulfur from coal. In Japan, stack gas scrubbers have worked successfully. And by the mid-1980s we can be converting coal to gas, extracting the sulfur in the conversion process and then piping the gas to individual homes for heating and cooling.

The development of nuclear energy poses a different type of environmental risk, centering on the safety of nuclear reactors and the safe custody of nuclear materials throughout the fuel cycle. Many persons, including myself, are therefore concerned about the projected growth of nuclear energy during the remainder of the century. My own feeling is that we need to redouble our efforts to build a greater safety factor into the design and operation of our nuclear plants and to insure greater protection from sabotage and from theft of material suitable for fabrication of nuclear weapons.

It is hoped that breakthroughs in solar energy by the twenty-first century may preclude the need for more fission reactions. In the meantime, we must not set a course which will require the construction of too many new reactors too quickly. The Half and Half Plan, for that reason, projects a total of 35 quadrillion BTUs of nuclear energy in 2000 as compared to 49.2 under the 1972 Department of Interior projection.

We must also focus more attention on the safeguarding and ultimate disposal of radioactive wastes . . . Currently, the mass of waste is still small, but even under the Half and Half Plan we will have hundreds of nuclear plants and therefore a great deal more waste products to deal with. We must therefore accelerate our efforts to develop more acceptable methods of handling this material.

A greater reliance on coal and nuclear energy means that we will be moving more and more to electricity in the residential and industrial sector. Oil supplies will have to be reserved to meet transportation needs and for petrochemical feed stocks. The use of greater amounts of electricity in turn means greater losses of energy in the process of converting fuel to power. In other words, we will have to burn more and more fuel to produce the energy that we can actually use. It will also require more transmission lines extending across our countryside.

As a result of the shock of the Arab boycott . . . [during the winter of 1973], there is considerable pressure these days to scuttle environmental constraints and go full-steam ahead to produce energy. Much of this pressure comes from those who fought the environmental movement in the first place and now see the present situation as an opportunity to win the battle they lost before. The Delaware coastal zone is an example.

When I was governor, Delaware already had extensive industrial areas and an existing refinery with room for expansion. We therefore decided that a 115-mile stretch of unspoiled marshes and beaches should not be developed for further energy facilities but should instead be preserved for their natural and recreational value. Now the pressures are building to reverse that decision. It would be a major mistake to give in. We need energy, but we need to protect our land.

It would be a mistake to trade an energy crisis for a health crisis.

The United States, then, must make a decision about its energy goals to the end of the century. We can attempt to continue the pattern of energy growth of the recent past, or we can make a serious commitment to energy conservation. In choosing between these two alternatives we need to reflect upon the impact of this choice on our lives.

The poor in our society can make good use of more energy. Meeting their needs is important, and the growth included in the Half and Half Plan can permit them to share more fully in the benefits conferred by our energy-based economy. But for most Americans, I find it difficult to conceive how a large increase in energy would improve the quality of life appreciably, even before considering the environmental costs and risks associated with this choice.

My vision of a future guided by the Half and Half Plan is of a nation undergoing some readjustments, but not fundamental ones. We would have to recycle our waste rather than continue to exploit virgin materials (recycling steel and paper, for example, requires 70 percent less energy), but this is manifestly desirable. We would have to become accustomed to smaller cars, and to walking more and driving less. We would be slightly cooler in the winter and slightly warmer in the summer, and we would turn off lights as we leave a room and consume fewer throwaway articles. Most important, we would all have to think about energy, as seriously as we now do about money. To my mind, few of these changes should provide any real discomfort.

Furthermore, a commitment to energy conservation may produce some unforeseen benefits. For example, as a continuing energy consciousness influences the planners and architects and citizens who create our homes and towns and cities, we are likely to find ourselves drawn more closely together. Instead of living in free-standing houses two miles from the store and twenty-five miles from work, we would move toward an intermixing of the various elements of our lives. The same will be true for our neighbors as well.

The end of profligacy in energy consumption may well force us to live with a greater sense of community—and we may just find that we like it.

A TIME TO CHOOSE [5]

A major energy study sponsored by the Ford Foundation recommends a government-led national commitment to conserve energy and argues that by such an undertaking

[5] From "50% Cut Proposed in Energy Growth," by Edward Cowan, staff reporter. New York *Times*. p 1. O. 18, '74. © 1974 by The New York Times Company. Reprinted by permission.

the United States can put off for ten years "massive new commitments" to offshore drilling, oil imports, nuclear power or development of coal and shale in arid areas of the West.

The final report of the foundation's Energy Policy Project . . . contends that the energy consumption growth rate can be cut by more than 50 percent without hurting the national economy.

"Energy growth and economic growth can be uncoupled," said S. David Freeman, the project director, at a news conference.

Among the report's key recommendations for conservation are the following:

☐ Higher energy prices, resulting from elimination of tax breaks for producers, such as the percentage depletion allowance, from enactment of pollution taxes and from charging consumers for "the costs of stockpiling oil" to protect the country against another embargo for foreign oil producers

☐ Federal assistance to low-income families, perhaps including "energy stamps" similar to food stamps or special payments or fuel allocations "to low-income persons who demonstrate potential hardship as a result of shortages or price increases for energy"

☐ Enactment by Congress of minimum gasoline economy standards for cars, with a goal of raising average fuel economy to twenty miles a gallon by 1985

☐ Federal loans to help householders and small businesses install insulation or energy-saving equipment

☐ Redesign of electricity rates "to eliminate promotional discounts and to reflect peak load costs"

Eventually, new energy sources will have to be tapped, the report says. But it asserts that a major conservative effort, including redirection of investment funds from energy supply to energy efficiency, would give the nation time to learn more about environment, health and economic risks of new supplies and make the wisest choice.

The report, entitled "A Time to Choose," is 343 pages long. . . .

The report, which with associated research cost the Ford Foundation $4 million, is considered a major contribution to the emerging national debate about how to meet the country's future energy needs.

Arguing that there is not much more time to debate and that it is now time to choose, the report says that the best thing to do at the outset is cut deeply into the country's appetite for energy, which has been exploding at a rate of 4.5 percent a year for eight years.

"Slower energy growth," the report says, "can work without undermining our standard of living and can also exert a powerful positive influence on environmental and other problems closely intertwined with energy," such as foreign relations.

The report was expected to add to pressures on the Ford Administration to adopt conservation actions stronger than the President's appeal . . . for voluntary restraint.

Mr. Freeman, when asked, said the Administration's policy was "grossly inadequate." . . .

The report expressed skepticism about public interest in energy issues. "It remains to be seen," the introductory chapter said, "whether citizens, in the absence of shortages, will sustain their interest in energy."

Traditionally, with energy cheap and abundant in the United States, "most citizens were content to let industry make the major decisions," the report continued. It then added:

"But the fact is that the private interests of energy companies and the broader public interest do not always coincide. What is good for business is not always good for the rest of the country."

The study contends that a so-called technical-fix scenario, or package of conservation actions, would slice the energy growth rate to 2 percent a year, by 1985, and that negligible growth thereafter could be achieved if such a policy goal were adopted.

Electricity growth would be somewhat higher than 2 percent, the report foresees, because "the nation is gradually moving towards a predominantly electric energy economy."

Nevertheless, the authors find, electricity growth would be about only half the customary rate of 7 percent. Consequently, "power plants now on order for completion by 1980 could satisfy the demand for electricity until 1985," making possible "a pause of several years in new power plant starts." . . .

Summarizing an extended discussion of safety problems associated with nuclear power plants, the report said:

"We do not advocate an absolute ban on new nuclear plants because the problems posed by using fossil fuels instead are also serious. But a conservation-oriented growth policy will provide breathing room so that we can gain a better understanding of nuclear power problems" before making "major new expansions."

Mr. Freeman said he personally was "nervously optimistic" that nuclear power plants would prove to be safe.

Other findings included these:

☐ "The United States energy-supplying industry does not constitute a monopoly by economic standards, although there are indications of diminished competition in some areas." The report recommended changes in offshore leasing policy to stop big companies from bidding jointly, and stronger antitrust enforcement.

☐ The Interior Department has made available to private interests "vast amounts of the resource base" with "a grossly inadequate return to the public treasury."

☐ Regulation of electric power rates by state regulatory commissions "is inadequate to cope with utilities that largely operate in regional power grids." Regional commissions "could assure that utility expansion plans were integrated into regional grids so as to meet regional needs with maximum efficiency."

☐ Because energy makes up a large and growing part of poor people's budgets, "the social equity implications of high energy prices should be resolved by a national commitment to income redistribution measures, such as a guaranteed minimum income or negative income tax."

IT TAKES ENERGY TO GET ENERGY[6]

In the mid-nineteenth century, a British company launched the *Great Eastern,* a coal-fired steamship designed to show the prowess of Britain's industrial might. The ship, weighing 19,000 tons and equipped with bunkers capable of holding 12,000 tons of coal, was to voyage to Australia and back without refueling. But it was soon discovered that to make the trip the ship would require 75 percent more coal than her coal-storage capacity—more coal, in fact, than the weight of the ship herself.

Today the United States is embarking on an effort to become independent in energy production, and such a program deserves the kind of analysis that the British shipbuilders overlooked. Indeed, our civilization appears to have reached a limit similar to that of the *Great Eastern:* The energy which for so long has driven our economy and altered our way of life is becoming scarce, and a number of respected experts are suggesting that, without significant changes, our society will go the way of the ship that needed more fuel than it could carry.

In recent years, energy growth in the United States has expanded at a rate of nearly 4 percent per year, resulting in a per capita consumption of all forms of energy higher than that of any other nation. US energy consumption in 1970 was half again as much as all of Western Europe's, even though Europe's population is one and a half times ours.

As energy consumption has increased in this nation, our

[6] From article by Wilson Clark, a consultant specializing in energy. *Smithsonian.* 5:84-90. D. '74. Copyright 1974 Smithsonian Institution, from *Smithsonian* magazine December 1974. Reprinted by permission.

energy resources have drastically declined. According to M. King Hubbert, a highly respected energy and resource expert, the peak for production of all kinds of liquid fossil-fuel resources (oil and natural gas) was reached in this country in 1970 when almost four billion barrels were produced. "The estimated time required to produce the middle 80 percent [of the known reserves of this resource]," Hubbert says, "is the 61-year period from 1939 to the year 2000, well under a human life span."

As available domestic oil and gas resources have declined, we have turned more and more to foreign imports—but, since 1973, the price of this essential imported oil has quadrupled. Recoiling from the specter of another embargo, federal officials and industrialists have suggested that the nation develop alternative energy sources such as nuclear power, and fossil fuels such as coal and oil shale, to bridge the energy gap and enable the nation to become self-sufficient.

According to John Sawhill, former chief of the Federal Energy Administration, "the repercussions of Project Independence will be felt throughout our economy. It will have a dramatic impact on the way 211 million Americans work and live." The price tag placed on pursuing the energy goals of Project Independence has been estimated to fall somewhere between $500 billion and $1 trillion. Raising such capital for energy development may prove to be the greatest financial undertaking in the history of the United States. A growing number of experts, however, say the goal of Project Independence may be unreachable.

The central problem is simply that *it takes energy to produce new energy*. In other words, in every process of energy conversion on earth, some energy is inevitably wasted. The laws of thermodynamics, formulated in the last century, might be viewed as describing a sort of "energy gravity" in the universe: energy constantly moves from hot to cold, from a higher to a lower level. Some energy is free for man's use—but it must be of high quality. Once used, it cannot be recycled to produce more power.

Coal, for example, can be burned in a power plant to produce steam for conversion into electric power. But the resulting ashes and waste heat cannot be collected and burned to produce yet more electricity. The quality of the energy in the ashes and heat is not high enough for further such use.

Numerous studies have indicated that the United States has enormous reserves of fossil fuels which can provide centuries of energy for an expanding economy, yet few take into account the thermodynamic limitations on mining the fuels left. Most cheap and accessible fossil-fuel deposits have already been exploited, and the energy required to fully exploit the rest may be equal to the energy contained in them. What is significant, and vital to our future, is the *net* energy of our fuel resources, not the *gross* energy. Net energy is what is left after the processing, concentrating and transporting of energy to consumers is subtracted from the gross energy of the resources in the ground.

Consider the drilling of oil wells. America's first oil well was drilled in Pennsylvania in 1859. From 1860 to 1870, the average depth at which oil was found was 300 feet. By 1900, the average find was at 1,000 feet. By 1927, it was 3,000 feet; today, it is 6,000 feet. Drilling deeper and deeper into the earth to find scattered oil deposits requires more and more energy. Think of the energy costs involved in building the trans-Alaska pipeline . . . For natural gas, the story is similar.

Dr. Earl Cook, dean of the College of Geosciences at Texas A. & M. University, points out that drilling a natural gas well doubles in cost each 3,600 feet. Until 1970, he says, all the natural gas found in Texas was no more than 10,000 feet underground, yet today the gas reserves are found at depths averaging 20,000 feet and deeper. Drilling a typical well less than a decade ago cost $100,000 but now the deeper wells each cost more than $1,000,000 to drill. As oil men move offshore and across the globe in their search for dwindling deposits of fossil fuels, financial costs increase, as do the basic energy costs of seeking the less concentrated fuel sources.

Although there is a good deal of oil and natural gas in the ground, the net energy—our share—is decreasing constantly.

The United States has deposits of coal estimated at 3.2 trillion tons, of which up to 400 billion tons may be recoverable—enough, some say, to supply this nation with coal for more than 1,000 years at present rates of energy consumption. And since we are dependent on energy in liquid and gaseous form (for such work as transportation, home and industrial heating), the energy industries and the Federal Energy Administration have proposed that our vast coal deposits be mined and then converted into gas and liquid fuels.

Yet the conversion of coal into other forms of energy, such as synthetic natural gas, requires not only energy but large quantities of water. In fact, a panel of the National Academy of Sciences recently reported that a critical water shortage exists in the western states, where extensive coal deposits are located. "Although we conclude that enough water is available for mining and rehabilitation at most sites," said the scientists, "not enough water exists for large-scale conversion of coal to other energy forms (*e.g.*, gasification or steam electric power). The potential environmental and social impacts of the use of this water for large-scale energy conversion projects would exceed by far the anticipated impact of mining alone." In fact, the energy and water limitations in the western states preclude more than a fraction of the seemingly great US coal deposits from ever being put to use for gasification or liquefaction.

The prospects for oil shale development are not as optimistic as some official predictions portend. Unlike oil, which can be pumped from the ground relatively easily and refined into useful products, oil shale is a sedimentary rock which contains kerogen, a solid, tarlike organic material.

Shale rock must be mined and heated in order to release oil from kerogen. The process of mining, heating and processing the oil shale requires so much energy that many experts believe that the net energy yield from shale will be negligible. According to *Business Week,* at least one major oil company has decided that the net energy yield from oil shale is so small that they will refuse to bid on federal lands containing deposits. And even if a major oil shale industry were to develop, water supplies would be as great a problem as for coal conversion, since the deposits are in water-starved western regions. The twin limiting factors of water and energy will preclude the substantial development of these industries.

Nuclear power is seen as the key to the future, yet an energy assessment of the nuclear fuel cycle indicates that the net energy from nuclear power may be more limited than the theoretically prodigious energy of the atom has promised.

Conventional nuclear fission power plants, which are fueled by uranium, contribute little more than 4 percent of the US electricity requirements at present, but according to the Atomic Energy Commission, fission will provide more than half of the nation's electricity by the end of the century. Several limitations may prevent this from occurring. One is the availability of uranium ore in this country for conversion to nuclear fuel. According to the United States Geological Survey, recoverable uranium resources amount to about 273,000 tons, which will supply the nuclear industry only up to the early 1980s. After that, we may well find ourselves bargaining for foreign uranium, much as we bargain for foreign oil today.

According to energy consultant E.J. Hoffman, however, an even greater problem with nuclear power is that the fuel production process is highly energy-intensive. "When all energy inputs are considered," he says, such as mining uranium ore, enriching nuclear fuel, and fabricating and operating power plants and reprocessing facilities, "the net electrical yield from fission is very low." Optimistic estimates from such sources as the President's Council on Environmental Quality say that nuclear fission yields about 12 percent of the energy value of the fuel as electricity: Hoffman's estimate is that it yields only 3 percent. That advanced reactors might have a higher net yield is one potential, but largely unknown at present, since such reactors have not yet been built and operated commercially. Other nuclear power processes, such as nuclear fusion, have simply not yet been shown to produce electricity, and so they cannot be counted upon. Even the more "natural" alternative energy sources, such as solar power, wind power and geothermal power, have not been evaluated from the net energy standpoint. They hold out great promise—especially from a localized, small-scale standpoint. Solar energy, for example, is enormous on a global scale but its effect varies from one place to another. However, the net energy yield from solar power overall might be low, requiring much energy to build elaborate concentrators and heat storage devices necessary.

What about hydrogen as a replacement fuel? By itself, hydrogen is not at all abundant in nature, and other energy sources must first be developed to power electrolyzers in order to break down water into hydrogen and oxygen. The energy losses inherent in such processes may result in a negligible overall energy yield by the time hydrogen is captured, stored and then burned as fuel. An indication of the magnitude of this problem has been given by Dr. Derek Gregory of the Institute of Gas Technology in Chicago, who points out that to substitute hydrogen fuel fully for the natural gas currently produced would require the construction of 1,000 enormous one-million-kilowatt capacity electric power plants to power electrolyzers—more than twice the present entire installed electrical plant capacity of the nation.

While much of this kind of analysis is apparently new to most energy planners, it also represents more than an analogy to the cost accounting that is familiar to businessmen investing dollars to achieve a net profit. The net energy approach might provide a new way of looking at subjects so seemingly disparate as the natural world and the economy.

Dollar Values of Natural Systems

An outspoken proponent of the net energy approach is Dr. Howard T. Odum, a systems ecologist at the University of Florida. In the 1950s, Odum analyzed the work of researchers trying to grow algae as a cheap source of fuel, and found that the energy required to build elaborate facilities and maintain algae cultures was greater than the energy yield of the algae when harvested for dry organic material. The laboratory experiment was subsidized, not by algae feeding on free solar energy—which might have yielded a net energy return—but by "the fossil-fuel culture through hundreds of dollars spent annually on laboratory equipment and services to keep a small number of algae in net yields."

With his associates at the University of Florida, Odum began to develop a symbolic energy language, using computer-modeling techniques, which relates energy flows in the natural environment to the energy flows of human technology.

Odum points out that natural sources of energy—solar radiation, the winds, flowing water and energy stored in plants and trees—have been treated as free "gifts" rather than physical energy resources which we can incorporate into our economic and environmental thinking. In his energy language, however, a dollar value is placed on all sources of energy—whether from the sun or petroleum. To produce each dollar in the economy requires energy—for example, to power industries. The buying power of the dollar, therefore, can be given an energy value. On the average, Odum calculates, the dollar is worth 25,000 calories (kilocalories, or large calories) of energy—the familiar energy equivalent dieters know well as food values. Of this figure, 17,000 calories is high-quality energy from fossil fuels and 8,000 calories low-quality energy from "natural" sources. In other words, the dollar will buy work equal to some mechanical labor, represented by fossil-fuel calories, and work done by natural systems and solar energy.

Odum's concept of energy as the basis of money is not new; a number of nineteenth century economists thought of money or wealth as deriving from energy in nature. The philosophy was expounded earlier in this century by Sir Frederick Soddy, the British scientist and Nobel laureate, who wrote that energy was the basis of wealth. "Men in the economic sense," he said, "exist solely by virtue of being able to draw on the energy of nature. . . . Wealth, in the economic sense of the physical requisites that enable and empower life, is still quite as much as of yore the product of the expenditure of energy or work."

Odum views natural systems as valuable converters and storage devices for the solar energy which triggers the life-creating process of photosynthesis. Even trees can be given a monetary value for the work they perform, such as air purification, prevention of soil erosion, cooling properties, holding ground water, and so on. In certain locations, he says, an acre of trees left in the natural state is worth more than $10,000 per year or more than $1 million over a hundred-year period, not counting inflation. Last year, he calculated that solar energy, in conjunction with winds, tides and natural ecological systems in the state of Florida, contributed a value of $3 billion to the state, compared to fossil-fuel purchases by the state's citizens of $18 billion per year.

The value of the natural systems to the state had never before been calculated. "These parts of the basis of our life," says Odum, "continue year after year, diminished however, when ecological lands that receive sun, winds, waves and rain are diverted to other use." He is now developing a "carrying capacity" plan for the future development of the state which has attracted the interest of the state legislature.

Odum's work may lead to eminently practical applications, by indicating directions in which our society can make the best use of energy sources and environmental planning. One application is to use natural systems for treating wastes, rather than using fossil fuels to run conventional waste-treatment plants. "There are," he says, "ecosystems capable of using and recycling wastes as a partner of the city without drain on the scarce fossil fuels. Soils take up carbon monoxide, forests absorb nutrients, swamps accept and regulate flood waters." He is currently involved in a three-year program in southern Florida to test the capability of swamps to treat wastes, and demonstrate their value to human civilization as a natural "power plant." The work, supported by the Rockefeller Foundation and the National Science Foundation, has drawn the attention and interest of many community and state governments.

According to Odum's energy concepts, a primary cause of inflation in this country and others is the pursuit of high economic growth with ever more costly fossil fuels and other energy sources. As we dig deeper in our search for less-concentrated energy supplies to fuel our economy, the actual value of our currency is lessening. "Because so much energy has to go immediately into the energy-getting process," he notes, "then the real work to society per unit of money is less."

Economists, who generally resent intruders on their turf, have not embraced this equation of energy and money with much enthusiasm, but it is gaining adherents in several quarters. According to Joel Schatz of Oregon's energy planning office, Odum's work leads the way toward effective government planning in this age of economic uncertainty. "The more successful the United States is in maintaining or increasing its total energy consumption," he says, "under conditions of declining net energy, the more rapidly inflation, unemployment and general economic instability will increase." Many people currently consider this disruption only an economic crisis, says Schatz, rather than what he believes it really is: a symptom of a continuing and deepening energy crisis.

There are signs that the net energy approach is being taken seriously even by the architects of Project Independence. Eric Zausner of the Federal Energy Administration says that net energy is a "useful concept" which is under investigation. "Net energy flows," he adds, "have practical implications in the new and exotic fuels, such as oil shale. With coal, there is no issue, since there is a net output of energy. But [for] some of the new processes, such as shale oil processing *in situ,* net energy flow is a very important consideration in whether we should do it or not."

Congressman George Brown Jr., a physicist from southern California and one of a bare handful of scientifically trained members of Congress, goes much further. He believes that the new Office of Technology Assessment in the Congress should undertake a broad energy analysis, encompassing the net energy approach, of the widespread implications of the Administration's plans for Project Independence. "We must start with the assumption that the energy available to do work is declining. This one assumption, which is firmly based on the laws of physics, will revolutionize economic policy once its truth becomes known. . . . The implications of the limits to growth of our economic systems are just beginning to be understood," says Congressman Brown, pointing out that the net energy approach indicates the inevitability of a national shift of emphasis toward a steady-state economy. "While this view is not yet widely held in Congress, the ranks of advocates are growing."

Since the Industrial Revolution, the Western world has been engaged in a great enterprise—the building of a highly complicated technological civilization. The Western "growth" economy (which today also characterizes Japan) has been made possible by seemingly endless supplies of inexpensive energy. One implication of the net energy approach is that a vigorous and wide-reaching conservation program may be the only palliative for inflation.

Another implication is that the days of high growth may be over sooner than most observers have previously thought. For it is increasingly apparent that today's energy crisis is pushing us toward a "steady-state" economy: No one yet knows what such an economy will look like or what social changes will result. But it would seem to be about time to start thinking seriously about it.

BIBLIOGRAPHY

An asterisk (*) preceding a reference indicates that an article or a part of it has been reprinted in this book.

Books and Pamphlets

Adelman, M. A. The world petroleum market. Johns Hopkins University Press. '72.

Berlin, Edward and others. Perspective on power: the regulation and pricing of electricity. (Ford Foundation Series) Ballinger. '74.

Boesch, D. F. and others. Oil spills and the marine environment. (Ford Foundation Series) Ballinger. '74.

Breyer, S. G. and MacAvoy, P. W. Energy regulation by the Federal Power Commission. Brookings Institution. 1775 Massachusetts Ave. N.W. Washington, D.C. 20036. '74.

* Citizen's Advisory Committee on Environmental Quality. Citizen action guide to energy conservation. Supt. of Docs. Washington, D.C. 20402. '73.

Connery, R. H. and Gilmour, R. S. eds. The national energy problem. Academy of Political Science. '74.

Davis, D. H. Energy politics. St. Martin. '74.

Dye, Lee. Blowout at platform A: the crisis that awakened a nation. Doubleday. '71.

Easton, R. O. Black tide: the Santa Barbara oil spill and its consequences. Delacorte Press. '72.

Engler, Robert. Politics of oil; a study of private power and democratic directions. University of Chicago Press. '67.

Fisher, J. C. Energy crisis in perspective. Wiley. '74.

Ford Foundation. Energy policy report. Lippincott. '75.

* Freeman, S. D. Energy: the new era. Walker; paper ed. Vintage. '74.

Fritsch, A. J. The contrasumers: a citizen's guide to resource conservation. Praeger. '74.

Garvey, Gerald. Energy, ecology, economy; a project of the Center of International Studies, Princeton University. Norton. '72.

Gordon, Howard, ed. Energy crisis handbook, Science Associates International, Inc. 23 E. 26th St. New York 10010. '74.

Gordon, Suzanne. Black Mesa: the angel of death. Day. '73.

* Holdren, John, and Herrara, Philip. Energy: a crisis in power. (Battlebooks Series) Sierra Club Books. '72. [distributed by Scribner]
Original title: Megawatt.
Reprinted in this book: Chapter 1, Understanding energy, by John Holdren; Chapter 3, Hydroelectric energy, by John Holdren.

Klebanoff, Shoshana. Middle East oil and U.S. foreign policy: with special reference to the U.S. energy crisis. Praeger. '74.

Knox, Susan. The energy crisis survival kit. Manor Books. '74.

Kruger, Paul and Otte, Carel, eds. Geothermal energy: resources, production, stimulation. Stanford University Press. '73.

Lawrence, R. M. and Wengert, N. I. The energy crisis: reality or myth. American Academy of Political and Social Science. 3937 Chestnut St. Philadelphia 19104. '73.

Lovejoy, W. F. and Homan, P. T. Economic aspects of oil conservation regulation. Johns Hopkins Press (for Resources for the Future, Inc.). '67.

Lovejoy, W. F. and Homan, P. T. Methods of estimating reserves of crude oil, natural gas, and natural gas liquids. Johns Hopkins Press (for Resources for the Future, Inc.). '65.

McDonald, S. L. Petroleum conservation in the United States: an economic analysis. Johns Hopkins Press. '71.

Mancke, R. B. Failure of U.S. energy policy. Columbia University Press. '74.

Marx, Wesley. Oilspill. (Battlebooks Series) Sierra. '72.

Megill, R. E. Exploration economics. Petroleum Publishing Co. P.O. Box 1260. Tulsa, Okla. 74101. '75.

Mikdashi, Z. M. The community of oil exporting countries: a study in governmental co-operation. Allen, G. '72.

Millard, Reed and the Editors of Science Book Associates. How will we meet the energy crisis? rev. ed. Messner. '74.

Miller, Leroy. The economics of energy: what went wrong? Morrow. '74.

Mosley, Leonard. Power play: oil in the Middle East. Random House. '74.

Mosley, Leonard. Power play: the tumultuous world of Middle East oil, 1890–1973. Weidenfeld. '73.

Oppenheimer, B. I. Oil and the congressional process. (Lexington Books) Heath. '74.

Organisation of Petroleum Exporting Countries. Annual statistical bulletin 1973. International Publications Service. 114 E. 32d St. New York 10016. '74.
Also available for 1967, 1968, 1971, and 1972.

Potter, Jeffrey. Disaster by oil; oil spills: why they happen, what they do, how we can end them. Macmillan. '73.

Pratt, W. E. and Good, Dorothy, eds. World geography of petroleum. Princeton University Press. '50.

Putnam, P. C. Energy in the future. Van Nostrand. '53.

Roberts, Keith, ed. Towards an energy policy. Sierra. '73.

Rocks, Lawrence and Runyon, R. P. The energy crisis. Crown. '72.

Rodgers, W. H. Brown-out, the power crisis in America. Stein & Day. '72.

Ruedisili, L. C. and Firebaugh, M. W. eds. Perspective on energy: issues, ideas and environmental dilemmas. Oxford University Press. '75.

Schurr, S. H. ed. Energy, economic growth, and the environment: papers presented at a forum conducted by Resources for the Future, Inc. in Washington, D.C., 20-21 April, 1971. Johns Hopkins University Press. '72.

Schurr, S. H. and Netschert, B. C. Energy in the American economy 1850–1975. Johns Hopkins Press (for Resources for the Future, Inc.). '60.

Scott, D. L. Pollution in the electric power industry. (Lexington Books) Heath. '73.

Shuttlesworth, D. E. and Williams, L. A. Disappearing energy: can we end the crisis? Doubleday. '74.

Shwadran, Benjamin. The Middle East, oil, and the great powers. (Halsted Press Book) 3d ed. rev. & enl. Wiley. '74.

Sobel, L. A. and others, eds. Energy crisis: 1969–73. Facts on File, Inc. 119 W. 57th St. New York 10019. '74.

Spies, H. R. and others. 350 ways to save energy (and money) in your home and car. Crown. '74.

Udall, S. L. and others. The energy balloon. McGraw-Hill. '74.

United States. Congress. House. Committee on Science and Astronautics. Subcommittee on Energy. Energy from oil shale: technical, environmental, economic, legislative, and policy aspects of an underdeveloped energy source: report prepared by the Science Policy Research Division, Library of Congress. 93d Congress, 1st session. U.S. Gov. Ptg. Office. Washington, D.C. 20401. '73.

* United States. Council on Environmental Quality. Energy and the environment—electric power. Supt. of Docs. Washington, D.C. 20402. '73.

Utton, A. E. ed. National petroleum policy: a critical review. University of New Mexico Press. '70.

Utton, A. E. and Henning, D. H. eds. Environmental policy: concepts and international implications. Praeger. '73.

Wilson, C. L. and Matthews, W. H. Man's impact on the global environment: assessment and recommendations for action. MIT Press. '70.

Periodicals

Annals of the American Academy of Political and Social Science. 410:24-34. N. '73. Future of American oil and natural gas. H. J. Frank and J. J. Schanz, Jr.

* Audubon. 76:81-8. My. '74. Pulling power out of thin air. Gary Soucie.

Barron's. p 3+. F. 10, '75. New lease on life: secondary recovery techniques are boosting production of crude. Armon Glenn.
* Bulletin of the Atomic Scientists. 31:4-5. Mr. '75. No alternative to nuclear power; 32 scientists speak out. H. A. Bethe and others.
* Business Week. p 80+. Ap. 27, '74. The looming oil battle off the East Coast.
Business Week. p 21-2. My. 4, '74. Big spending on oil, but no quick payoff.
Business Week. p 134+. My. 11, '74. The coal industry's controversial move west.
Business Week. p 82G+. My. 18, '74. The *in situ* way of turning coal into gas.
Business Week. p 114-17. O. 19, '74. Industry braces for a natural gas crisis.
* Business Week. p 38-44. F. 3, '75. The worldwide search for oil.
Business Week. p 66+. F. 10, '75. The case against both energy taxes and rationing.
Business Week. p 87-8. Ap. 28, '75. Shale oil's high-risk future.
Center Report. 8:20-4. Ap. '75. Prescription for the world petroleum problem. Neil Jacoby.
Challenge. 17:17-25. Ja./F. '75. Living with oil at $10 a barrel. Arnold Packer.
* Changing Times. 28:25-8. Jl. '74. Where else can we get energy?
Christian Century. 92:86-8. Ja. 29, '75. Public ownership of public utilities. R. K. Taylor.
* Commentary. 57:36-9. My. '74. Farewell to oil? E. N. Luttwak.
Commonweal. 99:527-30. Mr. 1, '74. Energy: end of an era. Michael Kennedy.
* Congressional Digest. 53:131-60. My. '74. Controversy over proposed U.S. regulation of surface mining of coal.
Congressional Quarterly Weekly Report. 31:3383-5. D. 22, '73. Shale oil development [on U.S. public lands]: a controversial venture.
Economist. 251:101-2. My. 4-10, '74. Sand and shale go boom.
Environment. 13:19-26+. D. '71. Power from the earth. David Fenner and Joseph Klarmann.
Environment. 14:14-22+. Ap. '72. Lost power. D. P. Grimmer and K. Luszczynski.
Environment. 14:29-33. My. '72. Fusion power. Lowell Wood and John Nuckolls.
Environment. 14:19-20+. Je. '72. When the well runs dry. R. H. Williams.
Environment Action Bulletin. v 4 [Special issue]. Je. 23, '73. Energy.

Environmental Action. 3:3-5. Ap. 1, '72. The breeder: abort it before it multiplies. Sam Love.
Esquire. 81:178+. My. '74. What's next in the oil crisis? Tad Szulc.
Forbes. 114:18-19+. D. 1, '74. The mote and the beam.
Forbes. 114:27+. D. 1, '74. How it all happened.
* Forbes. 115:57+. Ap. 1, '75. Shale oil: tantalizing, frustrating.
Fortune. 89:106-11+. Mr. '74. The far-reaching consequences of high-priced oil. Sanford Rose.
* Fortune. 89:104-7+. Ap. '74. Our vast, hidden oil resources. Sanford Rose.
Fortune. 89:214-19+. My. '74. Clearing the way for the new age of coal. E. K. Faltermayer.
Fortune. 89:136-9+. Je. '74. It's back to the pits for coal's new future. E. K. Faltermayer.
Fortune. 89:192-9. Je. '74. Oil, trade, and the dollar. L. A. Mayer.
* Fortune. 91:74-7+. Ja. '75. President Ford's hard choices on energy. Marilyn Wellemeyer.
National Geographic. 145:792-825. Je. '74. Oil, the dwindling treasure. Noel Grove.
National Parks and Conservation Magazine. 49:19-22. Ja. '75. Fusion-power to the people. J. L. Cecchi.
National Wildlife. 12:13-20. O-N. '74. Energy.
Natural History. 83:16+. O. '74. Short-circuiting the cheap power fantasy. G. M. Woodwell.
New Republic. 170:5-6. Mr. 9, '74. Shortchanging the environment.
New Republic. 171:17-19. S. 28, '74. Enough energy by 1985? Project Independence. Eliot Marshall.
New Republic. 171:5-7. O. 19, '74. Biting the bullet? Economic message of President Ford.
* New Republic. 172:17-19. F. 1, '75. The solar derby. Peter Barnes.
New Republic. 172:7-9. F. 15, '75. Holding back gas. Eliot Marshall.
New York Times. p 1+. F. 10, '74. Decades of inaction brought energy gap. Linda Charlton.
New York Times. p 58. Mr. 11, '74. White House challenged by environmental chief. E. W. Kenworthy.
New York Times. p 20. Jl. 14, '74. Environmental chief cautions on "panic" for "short-term energy gains." Gladwin Hill.
* New York Times. p 1+. O. 18, '74. 50% cut proposed in energy growth. Edward Cowan.
New York Times. p 1+. N. 13, '74. Mandatory saving is implicit in U.S. energy report. Edward Cowan.

* New York Times. p 68. N. 13, '74. Highlights of the [Project Independence] report.

New York Times. p 68. N. 13, '74. Missing: energy policy. Leonard Silk.

New York Times. p 27. Ja. 3, '75. The energy problem: what is to be done? P. G. Peterson.

New York Times. p 43. Mr. 4, '75. Qatar perturbed by cutback in oil. Eric Pace.

New York Times. p 13. Mr. 7, '75. 18 countries seek agreement on oil. C. H. Farnsworth.

* New York Times. p F 3. Mr. 9, '75. Focusing on energy—unclearly. J. W. Finney.

New York Times. p 20. Mr. 24, '75. Coal strip mining in West facing obstacles. B. A. Franklin.

New York Times. p K 26. Mr. 30, '75. Congress faces key decisions on nuclear reactors. David Burnham.

* New York Times. p 12. Ap. 2, '75. Energy and food needs clash in western states. Grace Lichtenstein.

New York Times. p E 6. Ap. 6, '75. The coming battle for the fortune at Elk Hills. Edward Cowan.

New York Times. p F 14. Ap. 20, '75. Uncle Sam would be a weak oil bargainer. J. H. Lichblau.

New York Times. p 25. My. 22, '75. Political axioms seen in energy impasse. D. E. Rosenbaum.

New York Times. p 1+. My. 28, '75. $1 oil import fee doubled by Ford; gas price to rise. J. M. Naughton.

New York Times. p 24. My. 29, '75. Houston, "energy capital" for world, ignores warnings of crisis. J. P. Sterba.

New York Times. p 32. My. 30, '75. U.N. assays geothermal energy.

New York Times. p 1. Jl. 1, '75. Energy development plan offers priorities for U. S. Edward Cowan.

New York Times. p 15. Jl. 29, '75. Nuclear power development encounters rising resistance with curbs sought in a number of states. Gladwin Hill.

New York Times Magazine. p 13+. D. 15, '74. Getting even; effect of oil prices on world monetary systems. Paul Lewis.

New York Times Magazine. p 10-11+. F. 16, '75. The economic political military solution. Daniel Yergin.

Newsweek. 83:117-18. My. 13, '74. Backsliding to oil; the collapsing coal-for-oil effort.

Newsweek. 83:58+. My. 27, '74. Laser energy.

Newsweek. 84:53-4. O. 14, '74. Can the oil cartel be broken

Newsweek. 85:23-4+. Ja. 27, '75. Energy: the price of saving: with views of Milton Friedman.

Newsweek. 85:76. F. 17, '75. The natural-gas gap. David Pauly and James Bishop, Jr.
Popular Science. 205:54-9+. Jl. '74. Wind power: how new technology is harnessing an age-old energy source. E. F. Lindsley.
Popular Science. 205:54-9+. Ag. '74. Fusion power: is it all coming together? Edward Edelson.
* Popular Science. 205:96-9+. N. '74. Geothermal energy . . . the prospects get hotter. J. F. Henahan.
Power. 119:24+. F. '75. Power from waste; special report. R. G. Schwieger.
* Progressive. 38:23-5. F. '74. Oil: the data shortage. Julius Duscha.
Progressive. 38:23-6. Mr. '74. Taking the lid off natural gas. S. E. Nordlinger.
Record (Bergen County Record, Hackensack, N.J.). p B-3. N. 25, '74. Offshore oil: the drawbacks. David Corcoran.
Record (Bergen County Sunday Record, Hackensack, N.J.). p D-1+. N. 24, '74. Offshore oil is coming and Jersey will never be the same. David Corcoran.
Saturday Review. 2:25-8. Ja. 25, '75. Whatever happened to Project Independence? S. E. Rolfe.
Saturday Review/World. 1:47-50. F. 9, '74. New energy sources. P. W. Quigg.
Science. 187:421+. F. 7, '75. Nuclear fusion: the next big step will be tokamak. W. D. Metz.
Science. 187:723+. F. 28, '75. Oil and gas resources: Academy calls USGS math "misleading." Robert Gillette.
Science: 187:795-803. Mr. 7, '75. Environmental impact of a geothermal plant. R. C. Axtmann.
Science and Public Affairs. 30:37-41. N. '74. Dilemma of fission power. D. P. Geesaman and D. E. Abrahamson.
Science and Public Affairs. 30:29-33. D. '74. Energy policy decision-making: the need for balanced input. Alexander De Volpi.
Science Digest. 76:32-7. O. '74. How to recycle a cow burp or some imaginative energy choices for the future. Douglas Colligan.
Scientific American. 230:19-20+. Mr. '74. The gasification of coal. Harry Perry.
Scientific American. 230:24-37. Je. '74. Fusion power by laser implosion. J. L. Emmett and others.
Scientific American. 232:34-44. Ja. '75. Fuel consumption of automobiles. J. R. Pierce.
Senior Scholastic. 104:8-31. F. 14, '74. Energy crisis [special issue].
Sierra Club Bulletin. 56:24-7. S. '71. Defusing Old Smoky by plugging into nature. John Holdren.
Sierra Club Bulletin. 57:16-18. F. '72. Electric power: an environmental dilemma.

Sierra Club Bulletin. 57:12-14+. S. '72. Strip mining: the biggest ripoff. Peter Borelli.
Sierra Club Bulletin. 58:10-14. Mr. '73. Cheap coal and hollow promises. William Greenburg.
Sierra Club Bulletin. 58:10-14+. My. '73. The realities and unrealities of energy economics. Mike Morrison.
Sierra Club Bulletin. 59:4-9. My. '74. Solar energy now. James Spaulding.
Smithsonian. 3:38-45. D. '72. Physicists probe the ultimate source of energy. Ben Bova.
* Smithsonian. 3:18-27. F. '73. Coal is cheap, hated, abundant, filthy, needed. Jane Stein.
Smithsonian. 4:70-8. N. '73. Interest in wind is picking up as fuels dwindle.
* Smithsonian. 5:80-5. Jl. '74. Nation's chief environmental adviser offers a long-range plan for energy. R. W. Peterson.
Smithsonian. 5:38-49. O. '74. Alaska embarks on its biggest boom as oil pipeline gets under way. Richard Corrigan.
* Smithsonian. 5:84-90. D. '74. It takes energy to get energy; the law of diminishing returns is in effect. Wilson Clark.
Sunset. 153:30-1. Ag. '74. Age of nuclear power is here, and even experts disagree about what's ahead.
Time. 101:41-2+. My. 7, '73. The energy crisis: time for action.
Time. 103:61-2. Ap. 1, '74. Power from gravity: use of black holes.
Time. 104:70-1. Ag. 12, '74. Move toward sharing: energy coordinating-group.
Time. 104:33-4+. O. 14, '74. Trying to cope with the looming crisis.
* Time. 105:8-28+. Ja. 6, '75. Faisal and oil.
U.S. News & World Report. 75:59-61. D. 24, '73. There's plenty of coal—what's behind the holdup?
U.S. News & World Report. 75:62. D. 24, '73. How practical is solar power?
U.S. News & World Report. 76:80-2. Mr. 11, '74. 250 billion barrels of oil just waiting to be "mined." Gerson Yalowitz.
* U.S. News & World Report. 76:43-4+. Je. 10, '74. Atomic power: a bright promise fading? Jack McWethy.
U.S. News & World Report. 77:23-4. O. 14, '74. Oil crisis: it threatens to bring down the West.
U.S. News & World Report. 77:22. D. 23, '74. Grave shortage of natural gas.
U.S. News & World Report. 77:38-40. D. 30, '74. Scramble for new oil—can it break the Arab strangle hold?
U.S. News & World Report. 78:37-9. Ja. 27, '75. Sun, sea, wind, geysers—new energy from old sources.

* U.S. News & World Report. 78:47-8. F. 3, '75. Natural-gas squeeze—how tight will it get?

Vital Speeches of the Day. 40:708-10. S. 15, '74. Natural gas shortage; address, August 20, 1974. J. C. Sawhill.

Vital Speeches of the Day. 41:9-13. O. 15, '74. U.S. energy supplies; address, September 12, 1974. W. P. Tavoulareas.

Wall Street Journal. p 1+. Ja. 4, '74. Shale sale: is shale oil an answer to energy shortages? D'Arcy O'Connor.

Wall Street Journal. p 1+. My. 3, '74. Energy alternative: gas and oil from coal can help overcome U.S. fuel shortage. Bob Arnold.

Wall Street Journal. p 1+. F. 28, '75. As an OPEC summit nears, price-shaving is more widespread. J. C. Tanner and James Carberry.

(7184)